INTRODUCTORY CHEMISTRY
In the LABORATORY

Third Edition

Bal Barot

M.S. (1982) University of Georgia
Ph.D. (1984) Oklahoma State University

TABLE OF CONTENTS

INTRODUCTION

Chemistry is everywhere and contributes to everyone's daily life; it has transformed our lives. Chemistry and chemists drive innovations to discover solutions to problems and enhance our lives (e.g. green chemistry). Chemistry is important to our Nation's economic health. It is vital that the next generation understands that chemistry is everywhere; is a transforming force that improves lives and our society; and is an important profession, because careers in chemistry offer opportunities to change the world.

Chemistry is an exciting and experimental science which lets us to understand our world and makes our life easier. Why soap cleans, but water and oil don't mix? Spinach, parsley and plants in general are green because they contain chlorophyll, a pigment which enables the plant to carry on photosynthesis, transforming solar energy and carbon dioxide into chemical energy in the form of carbohydrates and oxygen. Why colors of leaves on trees change in fall season? Why we are able to cook meals faster using pressure cookers and why cutting onions make you cry? Why smoking causes cancer? Nicotine is a naturally occurring liquid alkaloid. An alkaloid is an organic compound made out of carbon, hydrogen, nitrogen and sometimes oxygen. These chemicals have potent effects on the human body. For example, many people enjoy the stimulating effects of another alkaloid, caffeine. What is caffeine and how do chemists remove it to manufacture decaffeinated coffee? Why caffeine has such a strong effect on us? Caffeine operates using the mechanisms to stimulate the brain, though with stimulating effects. When you fall in love you may have many physical symptoms: loss of appetite, can't sleep, can't concentrate, sweat, butterflies in stomach. This is due to surging brain chemicals called organic chemicals. What is Organic Chemistry? Some people suffer nausea or diarrhea after drinking milk or milk derivatives. The origin of this problem can be the difficulty to digest lactose. What is lactose? When you go to the gas station, you choose gasoline of 92 octanes, or 89, or 87. What is octane? When you get sick, doctors prescribe antibiotics. What are antibiotics? Who is trained to make antibiotics? Besides being useful, chemicals find use in artifacts as well. Many materials like ceramics, paints, varnishes, glass, cement and various other useful substances contain chemicals as their components. What are these chemicals? Are they pure substances or mixtures? How do they differ from each other? Our body is made up of tissues, which are all composed of chemicals. We need an adequate supply of chemicals in the form of food, vitamins, hormones, and enzymes, which are chemicals. For taking care of our health, we need medicines.

We find that chemicals and chemistry penetrate into every aspect of our body, home and environment. Quality of life is improved through application of the principles of chemistry. So let us study chemistry to find answers to some of these questions.

Introduction for Laboratory

Grading Policy

Grading will be based on: your performances in the laboratory, lab reports, lab notebooks, answers to exercises, and any test for lab which may be given. The grading scheme is based on: Lab assignment, Write-ups, Clean working, and a written Lab final examination at the end of the term.

The technique grade reflects the student's preparedness for the lab, neatness, efficiency, ability to perform the experiments, ability to use the equipment properly, the product purity, and the product yield. Pay attention and cover assigned reading, procedures, and Lab Study Questions. The comprehensive lab final will be held in-class during the last lab period. Refer to Lab Notebook page for a description of what is to be included in the notebook.

Missed Labs

Unexcused absences will result in a grade of zero for the lab missed. No make-up labs will be allowed for unexcused absences. To have an absence excused, the student must:

- Phone the instructor before the lab starts.
- Email the instructor before the lab starts.
- Bring the proper documentations of the absence, e.g., a doctor's note.

Excused absences result in the averaging of the student's other lab grades and substituting this average grade for the lab missed. Missed labs must be made up if the student has missed more than 1 lab. To make up a lab, make arrangements before the end of the semester. Students who cannot meet these requirements must either drop the course or obtain permission and sign a contract to take an incomplete grade. Students who do not attend the first two laboratory sessions may be dropped from the lab and the lecture. You cannot drop the lab portion of chemistry without dropping the lecture, and vice versa. Do not assume you will be dropped for non-attendance; if you want to drop a class; it is your responsibility to drop it yourself. Please consult college student service for the procedure. Glassware Used for Chemistry must be clean. Rinse the glassware with the appropriate de-ionized water for water-soluble contents. If the glassware requires scrubbing, scrub with a brush using hot soapy water, rinse thoroughly with tap water, followed by rinses with de-ionized water. It is inadvisable to dry glassware with a paper towel or forced air since this can introduce fibers or impurities that can contaminate the solution. Normally you can allow glassware to air dry on the lab bench. Otherwise, if you are adding water to the glassware, it is fine to leave it wet (unless it will affect the concentration of the final solution).

LAB REPORT

At the end of the lab, you must submit a lab report. This science lab report includes pre-lab, experimental data report and answers to questions at the end of the lab experiment. It is important that you do pre-lab before coming to the lab.

Check-in:

Choose a workspace as soon as you come in the laboratory. Wait for instructions to get a tray from a cabinet. Once you get a tray, ask for a label. Write your name, semester and course name on the label. Place the label in the front metal frame of the tray. This is your tray for the semester, even though you may be allowed to work with a lab partner(s). Pull out the page from the lab manual which has Table-1, 'Student Tray Content'. Write your name and circle the items, which are missing in your tray from this list. Submit your check-in sheet to the instructor. Your drawer will be fully equipped before the end of the day.

Starting with Safety:

There are three parts to Starting with Safety.
Watch the video, take the safety quiz and sign the safety policy.
In chemistry laboratory, we must follow safety rules. It stands for occupational safety and health administration. The glassware and chemicals must be handled in a safe manner to avoid accidents. Watch the video carefully and learn safe ways to handle chemicals, thermometers, centrifuge, Bunsen burner and laboratory equipment. You must protect your eyes in the chemistry laboratory using protective safety goggles with side shields. These safety glasses are required and must be worn in the laboratory at all times. In the event of eye contacts chemicals, you should use eye fountain. Rinse your eyes for ten minutes and if still it hurts, you must see an eye doctor.
Wear sensible protective clothing in the laboratory. Avoid drinking or eating in the lab. Use small amounts of chemicals and if you have extra chemicals, do not return to the original container. Do not shake laboratory thermometers. Inform the instructor for any chemical spill. Wash glassware and your hands before you leave the laboratory. When you need to heat a liquid or solution, remember never heat a closed container or a flammable, volatile liquid using a Bunsen burner. In case of fire in the laboratory, follow instructions. Do not dispose chemicals in the sink. Always look for a labeled waste disposal container. If you have any question after you finish watching safety video, ask the instructor for clarification. Take the quiz and when you are done, let the instructor know it. The departmental policy asks you to read the safety policy, sign and write today's date. Give it to your instructor.

<u>**Pre-Lab Assignment 1**</u> **EXPERIMENT 1 Scientific Method**

Name:_________________________   **Date:**_________________

Q 1: Identify the following statements as either a hypothesis or theory or law:
(H = Hypothesis, T = Theory, L= Law)

1. All football players are fast runners.__________
2. All tall persons are great basketball players.______________
3. Chemistry problems can be solved using intuition.___________
4. The total mass of a matter remains same after a chemical change._____
5. All science students are excellent athletes.______________
6. Only if you are rich, then you can be happy.______________.

Q. 2: Provide the correct answer:

a) When a hypothesis is proposed, it is to _______________________________

b) Give an example of hypothesis

____

____

c) Theory is proposed by the scientist for

____

____

d) You leave your residence to come to college, and your car does not start, propose the steps using scientific method to solve the problem.

Q 4: Identify the following statements as either a hypothesis or theory or law or wrong:
(H = Hypothesis, T = Theory, L= Law, W=Wrong)
1. All students are honest.___________
2. All Chemistry problems involve use of computers.____________
3. The total volume of chemical substances remains same after a chemical change.______
4. All students are commuters.__________________
5. If you work hard, you can be happier.__________________.

Q. 2 Circle the correct answer:

1. A theory is
a) Brief generalization of many facts in a statement
b) An educated guess from the past experiences
c) A hypothesis that has been validated
d) Tentative explanation of an expert

2. All of the statements below are hypothesis except
a) All American students are excellent in health.
b) All college students are young.
c) People go to places for fun and entertainment.
d) Chemistry is a science, which involves study of properties of matter.

3. All of the statements below represent either a scientific law except
a) The total mass of a matter remains same after a chemical change
b) Energy of the universe remains constant.
c) The earth is round.
d) Tall trees live shorter time than short trees.

EXPERIMENT 1 Scientific Method
<u>Introduction:</u>

The way a scientist solves a problem is simple. A systematic and stepwise method called "The Scientific Method" is applied in solving research problems. In a chemistry laboratory, all experiments involve the application of the scientific method.

The scientific method involves data collection: When you measure and record the amount of water that is one set of data. If you heat that water and determine the temperature of the hot water that is another set of data. If you take a substance and measure and record its weight before adding it in hot water that is also a set of data. All these observations provide data.

The scientific method is done to solve a problem. The scientific method is a way to solve scientific questions by making observations and doing experiments. The steps of the scientific method are to:

<table>
<tr><td>1.</td><td>Define the problem by asking questions and making observations</td></tr>
<tr><td>2.</td><td>Do Background Research by collecting al relevant information</td></tr>
<tr><td>3.</td><td>Construct a Hypothesis to explain and offer a potential solution</td></tr>
<tr><td>4.</td><td>Test Your Hypothesis by Doing an Experiment</td></tr>
<tr><td>5.</td><td>Analyze Your Data of the experiment and draw a Conclusion</td></tr>
<tr><td>6.</td><td>Communicate Your Results as relate to your hypothesis</td></tr>
</table>

It is important for your experiment to be an experimental test. A "fair experimental test" occurs when you change only one factor (variable) and keep all other conditions the same. Let us say you want to study the effect of sunlight on plant growth. So your question is: Does sunlight promote plant growth? You must keep one plant in dark and another identical plant exposed to sunlight. All other factors like amount and frequency of water, fertilizer must be kept constant. A <u>hypothesis</u> is an educated guess or proposition that attempts to explain the reason and solution for the problem. And then the <u>scientific method</u> is used to test it. A hypothesis answers "why" a problem occurs, but a theory can explain "how" it takes place.

Examples of Hypotheses: **1.** If the water faucet is opened more, then the amount of water flowing will increase. **2.** If a prisoner learns a work skill while in jail, then he is less likely to commit a crime when he is released and will be a productive citizen in the future. **3.** If I raise the temperature of a cup of water, then the amount of sugar that can be dissolved in it will be increased. **4.** If there is a relation between the wealth and happiness, then the more wealth you collect, happier you become. **5.** If temperature is related to the rate of dissolving sugar, then raising the ambient temperature will cause an increase in sugar dissolution.

6. More students get sick during the winter than at other times. **7.** One solar panel on the house roof can save monthly utility bill. **8.** There is a positive correlation between job satisfaction and worker turnover. **9.** Worker satisfaction increases worker productivity. **10.** Amount of sun exposure will increase the growth of a plant.**11.** Childhood obesity is tied to the amount of sugary drinks taken daily. **12.** A dog can be trained to smell narcotic drugs.

Regardless of the type of hypothesis, the goal of a hypothesis is to help explain the problem. After data collection, a problem can be explained by an educated guess. Estimation is the first step to solve a problem. It may or may not be correct. It gives a ballpark figure, which may be good enough in the beginning. For example, if you are not feeling well, that is a problem. The first step to solve this problem is to collect data and observations.

As such, a hypothesis will:

- State the purpose of the research, how will it solve the problem
- Identify what variables are used in the experiment

In order to be a good hypothesis that can be tested or studied, a hypothesis must be logical and should be testable with research or experimentation. A hypothesis is usually used in a form where it proposes that if something is done, then something else will occur.

There are three types of scientific statements: Hypothesis, Law and Theory.

A hypothesis will give a plausible explanation that will be tested. It can be tested. A scientific theory is broader in scope and explains. A scientific theory often seeks to synthesize a body of evidence or observations of a particular problem. You can't necessarily reduce a scientific theory to a scientific equation, but it does represent something fundamental about how nature works.

Both laws and theories depend on basic elements of the scientific method, such as generating a hypothesis, testing by doing experiments and collecting evidence and coming up with conclusions. Eventually, other scientists must be able to replicate the results. Let us look at a common problem.

Hypothesis may not be correct every time and if the problem persists, then such hypothesis should be discarded. More observations and further data collection should help to develop a better and new hypothesis.

<u>Problem:</u>   Not feeling well. What should be the steps using a scientific method?

1. Take your body temperature, using a modern thermometer and find out if you have fever or not.
2. If you are sneezing, coughing and having a headache, those are considered observations. At this point you may try to guess and offer explanation for your problem.
3. Let's say it's the flu. This is called a hypothesis. How can you prove that your hypothesis is correct?
4. An experiment can provide the answer. If it is the flu, drinking plenty fluids, taking vitamin-C and resting for few days, should solve the problem.
5. If it does, then your hypothesis is correct.

Theory:

A theory proves pattern recognition and has the support of past data. It provides an explanation to the problem. However an experiment or a series of experiments are needed to develop a theory. Each experiment may involve a unique strategy. Some experiments require having a control and variables, while other experiments involve repetition of data collection, tabulation and pattern recognition.

Examples:

1. Usually, it is cold in winter and hot in summer.
2. The air on a mountain is pure.
3. The cities are crowded.

A hypothesis is an educated guess. It is based on a person's past experiences, who offers a tentative explanation to a problem. A theory proves pattern recognition and has the support of data. The main difference between these two terms is the fact that a hypothesis answers "why" a problem occurs, but a theory can explain "how" it takes place.

<u>**Scientific Law**</u>:

The scientific law is a true statement, which summarizes many observations in a concise and efficient manner. In 1687, Isaac Newton saw an apple falling from an apple tree and developed the law of gravity. It is still true today, without exception. The application of the scientific method empowers a scientist to predict. The power of prediction helps the development of technology. From the law of gravity, one can predict that if you throw a book or banana, it'll fall down toward the earth and will not go up in the sky. Examples of scientific laws: The oil and water don't mix. When a substance burns, its temperature increases.

EXPERIMENT 1 Scientific Method
<u>Procedure:</u>

Name:____ **Date:**_______________

<u>Objective:</u>

1. Learn the scientific method by proposing a hypothesis.
2. Learn to do simple experiment, using a separatory funnel.
3. Learn to observe and collect data to verify your hypothesis.

<u>Procedures</u> – Make a detailed list of the steps in your experiment.

Step 1: Take water in a beaker and heat it to 80^0 C using hot plate. Add blue food color to it. Transfer, carefully wearing hot gloves, 25 mL of it to a separatory funnel, kept in the ring stand, as shown here:

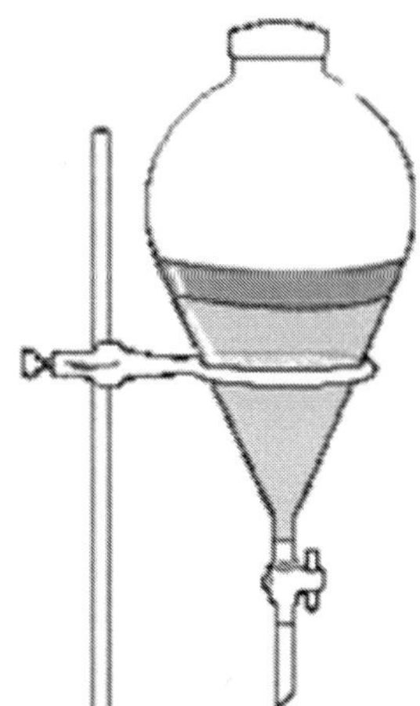

Step 2: Take cold water in another beaker. If needed, add ice to it. Record the temperature. It should be less than $10\ ^0$C. Add green food color to it. Transfer 25 mL of it to the separatory funnel which contains hot blue water, slowly and carefully.
Step3: Record your observations.
Step 4: Reverse the order. First take 25 mL of cold green water in a clean graduate cylinder. Transfer it to separatory funnel and then add 25 mL of hot blue water to it. Observe and record your observations.

<u>**Experiment 1**</u>

Name: ________________   Date:_________________________________

<u>**Lab Report:**</u>

1. Problem – What are you trying to figure out?

____

2. Your Hypothesis: What do you think will happen?

____
____
____

<u>**3.**</u> What did you observe when you performed the experiment?

a) Do they mix?_________________________________
b) Does hot blue water remains at the bottom layer?_________________
c) Does cold green water go to the bottom?_______________________

<u>**Conclusion**</u> – From what you observed, how would you answer your original question? Explain your results.

Experiment 2 Measurements in Metric System and Density Calculations

Pre-Lab Assignment

Name:_______________________   **Date:**___________________

1. Perform [Metric to Metric] Measurement Conversions:

1.   3.68 kg = ___________g

2.   568 cm = ___________m

3.   8700 ml = ___________l

4.   25 mg = ___________g

5.   0.101 cm = ___________mm

6.   250 ml = ___________l

7.   600 g = ___________kg

8.   8900 mm = ___________m

2. Perform Metric to English Measurement Conversions

1.   74 cm = ___________in.

2.   50 kg = ___________lbs.

3.   220 kg=___________________oz.

4.   160 km = ___________mi.

5.   3.6 lit = ___________gal.

6.   500 g = ___________oz.

7.   100 m = ___________yds.

8.   600 g = ___________lbs.

3. Read each sentence and classify it as either QL (Qualitative) or QT (Quantitative):

1. Table salt and sugar are solid substances.
   QL   QT
2. The ring was shiny and round.
   QL   QT
3. Loretta is a beautiful woman.
   QL   QT
4. Chemistry is a difficult and hard subject.
   QL   QT
5. The B S Inc. is doing well.
   QL   QT
6. Add 2.0 L of soft drink in the ice cream to make punch for the party.
   QL   QT
7. I lost 1.0-kg bodyweight after attending the camp.
   QL   QT
8. Sue Johnson is 6 feet tall.
   QL   QT
9. Poor people do not own wealth.
   QL   QT

4. Count significant figures in each case and write the correct answer:
1. 5000.2 _______________
2. 408_______________
3. 0.00056_______________
4. 4000_______________
5. 20.090_______________

5. Fill in the blanks:

a. A block of aluminum occupies a volume of 15.0 mL and weighs 40.5 g. What is its density?____________________

b. Mercury metal is poured into a graduated cylinder that holds exactly 22.5 mL. The mercury used to fill the cylinder weighs 306.0 g. From this information, calculate the density of mercury.________

c. What is the weight of the ethyl alcohol that exactly fills a 200.0 mL container? The density of ethyl alcohol is 0.789 g/mL.____________

d. A rectangular block of copper metal weighs 1896 g. The dimensions of the block are 8.4 cm by 5.5 cm by 4.6 cm. From this data, what is the density of copper?__________________

e. The density of sulfuric acid if 35.4 mL of the acid weighs 65.14 grams is_______

f. The mass of 250.0 mL of benzene with density of 0.8786 g/cc is______

Experiment 2 Measurements in Metric System and Density Calculations

Objectives:

1. To learn about the metric system and units used in length, volume, temperature and mass.
2. To become familiar with laboratory glassware and instruments like meter stick, burette, graduated cylinder, Erlenmeyer flask and balance.
3. To learn about the concept of density and determine the density of solid aluminum and liquid distilled water.
4. To express the results of the experiment in terms of the percentage error.
5. To find the density of unknown solid and liquid.

Introduction:

Using a table of conversions, you should be able to covert mass, volume, distance and temperature from one system to another. Whenever a very large or a very small number is given, you can avoid writing too many zeros and express them as a product of a number between one and ten and an exponential power of ten. The use of a calculator results in many numbers. So understanding rules of significant figures in your calculations will help you to simplify numbers.

Significant Figures:

The term significant figure is used to describe measured numbers plus one more, to show uncertainty in the measurement. This may be due to the instrument or measuring device used.

Consider the following situations. There are two rulers and two distinct and different measurements.

Picture

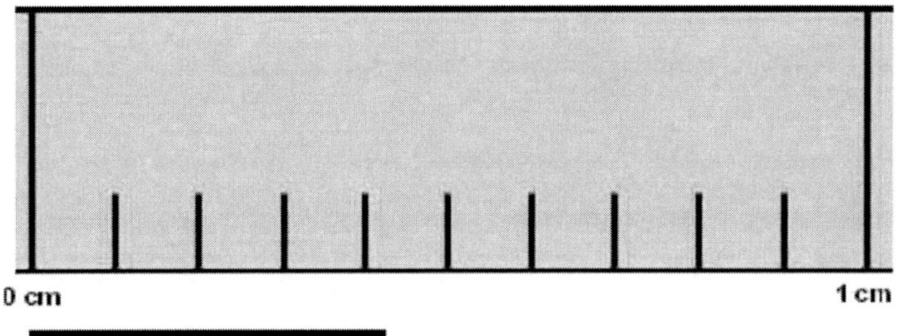

Uncertainty in Measurement

In the first case, there are no marks between 0.4 cm and 0.5, so how do we decide where the level stops. If one says it is 0.42 and another person says it is 0.44, the two answers are not identical. But both people are correct. In the second case, due to markings, one can say the arrow stops at 5.35 cm or one can say it goes up to 5.36 cm. Again two different answers and both of them are correct. When you record a measurement with certain number plus one more estimated number, it is you are recording significant figures.

The term significant figure is used to describe measured numbers plus one more, to show uncertainty in the measurement. This may be due to the instrument or measuring device used. Consider the following situations. There are two rulers and two distinct and different measurements. When you record a measurement with certain number plus one more estimated number, it is you are recording significant figures.

Rules of Significant Figures:

Non-zero numbers like 1,2,3,4 and so on are always significant.
There are three types of zeros:

*a)   **Captive**: These are the zeros in the middle, like 203.6 or 101 and they are always significant digits.*

*b)   **Leading**: The zeros in the beginning of a non-zero number are not significant. For example, 0.09865 has two zeros in the front of 9865, but they are not significant.*

*c)   **Trailing**: These zeros may or may not be significant.*

3. Significant trailing zeros: In 230.0 number, both zero after number 3 are significant.

Non-significant trailing zeros: In 80 numbers, there is only one significant digit and that is number 8. Zero after number 8 is not considered as significant digit.

Metric and English System:

We use the English system in the USA. However the rest of the world is using a different system, called the Metric System or S.I. The Metric System of weights and measurements came into existence in seventeenth century France. It allowed scientists from all over Europe to communicate freely with each other. In 1968, the US congress passed a bill recommending the change to the Metric System. Now the new system is called SI or International System, which is based on the same metric units. We will use the metric system in chemistry. The four fundamental units are kilogram for the mass, second for the time, meter for the length and Kelvin for the temperature.

METRIC SYSTEM

LENGTH

Unit	Abbreviation	Number of Meters	Approximate U.S. Equivalent
kilometer	km	1,000	0.62 mile
meter	m	1	39.37 inches
decimeter	dm	0.1	3.94 inches
centimeter	cm	0.01	0.39 inch
millimeter	mm	0.001	0.039 inch

VOLUME

1 quart = 0.946 liters   1 liter = 1.06 quart
1 gallon = 3.79 liters   1 liter = 0.264 gallons

MASS AND WEIGHT

kilogram	kg	1,000
gram	g	1
decigram	dg	0.10
centigram	cg	0.01
milligram	mg	0.001
microgram	μg	0.000001

Volume:

The standard unit of the volume we use in the Metric System is liter. One liter equals 1000 mL. One mL is the same volume as in one-centimeter cubed. Typical laboratory glassware has measuring units of mL.

<u>Beaker and Erlenmeyer flask are used to measure the volume of liquid.</u>

To understand the relationship between liters and gallons, you should take an empty milk gallon container and two coke bottles (2.0-Liter size). You will need to have two 2.0L soft drink bottles, with water and now if you transfer all the water from both bottles to the empty gallon container, very little will be left over.

One gallon equals to approximately 4.0 L, or to be accurate 3.78 L.

<u>Example:</u>

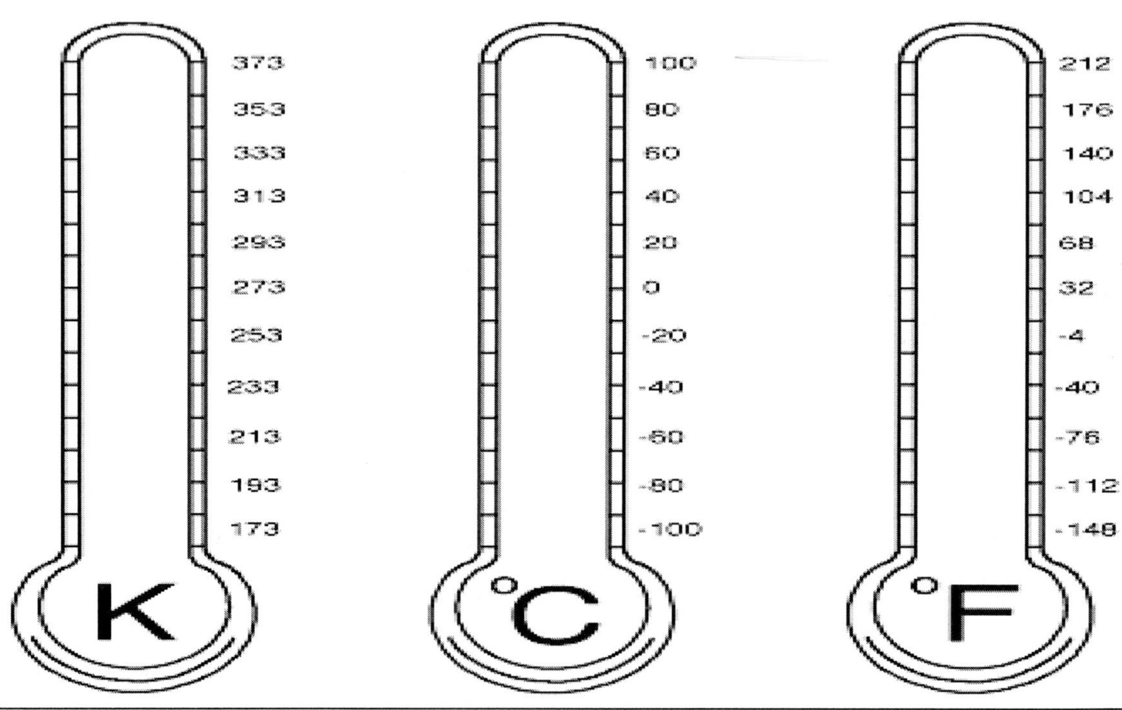

Temperature Scales:

We record temperatures in the USA in Fahrenheit. For example, water freezes at 32^0 F and boils at 212^0 F. Gabriel Fahrenheit, a German Scientist, devised this scale, in 1724.Anders Celsius, a Swedish scientist, later offered Celsius scale.

By definition, the freezing point of water in this scale is 0^0 C and boiling temperature of water is 100^0 C. The third type of temperature scale is called an absolute temperature scale and the temperature is measured in Kelvin scale. Theoretically, the lowest possible temperature in this scale is zero Kelvin (K). According to this scale, the temperature at which water freezes is 273 K and boiling point of water is 373 K.

Density is mass per unit volume. Use Handbook of Chemistry and Physics and find the literature value of density of solid aluminum and pure water at room temperature. Now calculate the error by finding the difference between found value of density and accurate value from the handbook. Convert the error in percentage. If it is more than 5%, you should repeat the experiment, before measuring the density of unknown compound.

$$\text{Density} = \frac{\text{Mass in grams}}{\text{Volume in ml}}$$

1. _Calculate the density of a sample in the organic laboratory. The weight of the sample was found to be 169 grams and its volume was 110 ml._

Solution:

Density = Mass/ Volume

Mass = 169 grams

Volume = 110 ml

Now substitute these values in the above formula of the density:

Density =Mass/ Volume

$$\text{Density} = \frac{169 \text{ grams}}{110 \text{ ml}} = 1.54 \text{ g/ml}$$

2. _Find the volume of a sample of urine collected for drug testing with a mass of 10.0 grams and 1.050 g/mL density._

Solution:

The sample mass is 10.0 grams. Density is 1.050 g/ml. The density is

$$\frac{\text{Mass}}{\text{Volume}} = \frac{10.0 \text{ g}}{\text{Volume}} = 1.050 \text{ g/ml}$$

$$\frac{10.0 \text{ g}}{1.050 \text{ g/ml}} = \text{Volume}$$

=9.52 ml

3. If a sample of alcohol has density of 0.789 g/mL and the volume of 100.0 mL, what will be its weight?

Solution:

Density   x   Volume   =   Mass

0.789 g/ml   x   100.0 ml   =   Mass

78.9 Grams.

Lab Exercise #2 Measurements in Metric System and Density Calculations

Procedure:

1. Measuring Length:

Use a meter stick, measure your and your lab partner's height. Record each height in the following units:

Meters
Centimeters
Millimeters

2. a. Reading volume:

Look at the meniscus of the liquid levels in the three graduated cylinders and record the following observations:

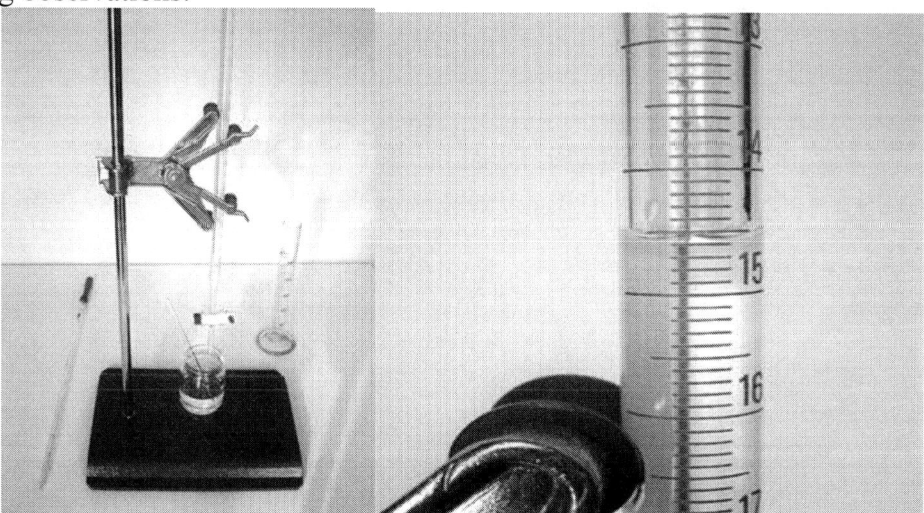

a)   Level in small graduate cylinder
b)   Level in medium graduate cylinder
c)   Level in large graduate cylinder

b. *Measuring volume*

i)   Using a beaker, take 25 ml of water.
ii)   Transfer it to small Erlenmeyer flask. Record the level in Erlenmeyer flask in ml.
iii)   Then transfer it to the graduate cylinder which is provided in your tray. Check the level and record your observation.
iv)   Now pour this water in the buret. Check the level and record your observation.

Measuring temperature:

Fill the large beaker with ice and water and stir with the stirring rod for two minutes. Keep the thermometer inside the beaker for two minutes and then raise it. Measure the temperature and record.

Measuring mass:

This measurement is very important for success in the laboratory. In the laboratory, the mass must be measured in metric units. The instructor will demonstrate how to use the balance, but you must obey the following guidelines:

When nothing is on the balance, it must read zero.
ii)   If it does not show zero reading, clean the balance pan.
iii)   Do not weigh anything directly on the pan. Always use a beaker, evaporating dish, watch glass, plastic boat or aluminum foil.
Do not any object when it is hot. Let it cool to room temperature.
Always record the weight, before you leave the balance room.
Clean the pan, in case if you spill anything.

4 (a) Using the balance first weigh the evaporating dish. It is in your tray. Record the mass.

4 (b) Weigh the mass of the medicine dropper and record.

4 (c) Keep the medicine dropper in the evaporating dish and place it on the pan and measure the combined mass. Record it.

Add the mass from procedure step 4(a) and 4 (b), if it differs from 4(c) by 0.030g, consult the instructor.

<u>Now for density determination:</u>

5(A) Density of Aluminum solid:

Choose an aluminum solid metal block and record its unknown number. *5(A) a.* Determine the mass and record *5(A)b.* Fill up the graduate cylinder with 20 ml of water. Record the initial level *5(A) c.* Place the aluminum block in the cylinder, by carefully sliding it inside. Record the new level. *5(A) d.* Calculate the difference between final volume of water and initial level of water in the graduate cylinder. That is the volume of the solid aluminum, record in *5(A) e.* Use the formula of density and by substituting mass in grams and volume in ml, calculate the density of aluminum, and record it, *5(A) f.* Now look at the accurate value of aluminum. Find your error by calculating the difference. Now compute percentage error. If it is more than 5%, repeat the experiment.

5(B) Density of unknown solid:
1. Choose a solid metal block and record its unknown number. *5(B) 1.*
2. Determine the mass and record. *5(B)2.*
3. Fill up the graduate cylinder with 20 ml of water. Record the initial level. *5(B) 3.*

4. Place the metal block in the cylinder, by carefully sliding it inside. Record the new level. *5(B) 4.*

5. Calculate the difference between final volume of water and initial level of water in the graduate cylinder. That is the volume of the solid metal. Record it. *5(B)5*

6. Use the formula of density and by substituting mass in grams and volume; calculate the density of unknown metal. Record it in *5(B) 6.*

Check with the master list and identify the probable solid and record in *5 (b) 7.*

5(C): Density of distilled water:

1. First take 25 ml of distilled water using pipet. Record the volume. *5(C):1.*
2. Weigh a clean and dry small beaker. Record the mass. *5(C):2.*
3. Now add 25 ml of distilled water to this beaker. Record the combined mass. *5(C):3.*
4. Calculate the mass of the liquid distilled water by finding the difference between combined mass and the mass of beaker only. *5(C):4.*
5. Compute the density of distilled water by dividing the mass of it in grams by its volume in ml. *5(C):5.*
6. Take the temperature and record it in *5(C):6.*
7. Find the true density from table at the end of this activity. *5(C): 7________*
8. Calculate the error. *5(C):8______________* and % error, record in *5C):9.*

5(D): Density of unknown solution:

1. First record the unknown number of the liquid solution. (5D1)

2. Take 20 ml of it as accurately as possible using pipet. Record the volume. (5D2)

3. Weigh a clean and dry small beaker. Record the mass. (5D3)

4. Now add 20 ml of the unknown liquid to this beaker. Record the combined mass. (5D4)

5. Calculate the mass of the unknown liquid by finding the difference between combined mass and the mass of beaker only. (5D5)

6. Compute the density of unknown liquid by dividing the mass of it in grams by its volume in ml. (5D6). You do not need to calculate % error. But see if you can identify the unknown liquid and name it. (5D7)

Lab Exercise Report #2 Measurements in Metric System and Density Calculations
Name____Date:_____________________________

Experimental Data:

1. Measuring Length:

Your height:

a)____________m
b)__________cm
c)____________mm

Your lab partner's height:

a)______________m
b)______________cm
c)______________ mm

2a): *Reading volume:*

Small graduate_______ml
Medium grduate______ml
Large graduate_______ml

2b): *Measuring volume:*

Level in beaker:_________ml
Level in small Erlenmeyer_______ ml
Level in the graduate cylinder _______________ml
Level in the buret _______________ml

Measuring temperature:

Temperature in beaker: ___________C, ___________F and ___________K

4. Measuring Mass:

Mass of evaporating dish   _________g
Mass of medicine dropper   _________g
Mass of evaporating dish   +   medicine dropper   _______g
% error:___________________________

5(A) Density of solid aluminum:

   a. Unknown number: _______________
   b. Mass of solid aluminum: _______ g
   c. Initial volume in graduate: _______ml
   d. Final volume in graduate: _______ ml
   e. Volume of solid aluminum: _______ ml
   f. Density of aluminum: _________g/ml

Mass of aluminum in g

Density= --------------------------------- =   g/ml
          Volume of aluminum in ml   _____________

Accurate density of Aluminum =2.702 g/ml

Error= Difference between 2.702 and experimental value from 3(A) f=_________________

          Error

% error = ----------------   x 100   =_________________________________

          Accurate

5(B) Density of solid unknown:

   1. Unknown number: _____________
   2. Mass of solid unknown: _______ g
   3. Initial volume in graduate: _______ml
   4. Final volume in graduate: _______ ml
   5. Volume of solid unknown: _______ ml
   6. Density of unknown: _________g/ml
   7. Identity of unknown solid_____________

          Mass of unknown in g

Density= ---------------------------------   =   g/ml
          Volume of unknown in ml   _____________________

5(c) Density of distilled water:

   1. Volume of distilled water: _______ ml
   2. Mass of empty beaker: _________ g
   3. Mass of beaker and distilled water=_______ g

4. Mass of distilled water=__________ g
5. Density of distilled water: __________g/ml

$$Density= \frac{Mass\ of\ distilled\ water\ in\ g}{Volume\ of\ distilled\ water\ in\ ml} = \underline{\hspace{3cm}}\ g/ml$$

6. Distilled water temperature __________ ^0C
7. True density from table__________________
8. Error:__________________
9. % error = -------   x 100   = __________________

5(D) Density of unknown liquid solution:

1. Unknown number: ______________
2. Volume of liquid unknown: ______ml
3. Initial mass of empty beaker: ______g
4. Final combined mass of beaker and unknown liquid: ________g
5. Mass of unknown liquid: ________g
6. Density of unknown liquid: __________g/ml
7. Identity of unknown liquid:__________________

POST-LAB Questions:

1. Who is taller, you or your laboratory partner? How many decimeters?

2. Do all the three measuring devices used to measure the volume give the volume in same number of significant figures? Which one would you use for precise measurement of volume?

3. Why do we use a mixture of ice and water, rather than only ice to measure the temperature?

4. Did you use the same balance during the mass determination of evaporating dish and medicine dropper? Why?

5. Identify the glassware you used today, draw them and label all of them.

6. Multiple choices and choose one correct answer:

a. Calculate the <u>density</u> of a metal, which has a volume of 5.00 cm^3 and it, is weighing 15.0 grams. It should be
a) 30 g/cc
b) 3.0 g/cc
c) 0.3 g/cc
d) none of the above

b. Find the <u>volume</u> of a sample of urine collected for drug testing with a mass of 10.0 grams and 1.050 g/mL density.
a) 9.52 ml
b) 95.2 ml
c) 952 ml
d) none of the above

c. If a sample of alcohol has density of 0.789 g/mL and the volume of 100.0 mL, what will be its <u>weight</u>?
a) 0.789 g
b) 7.89 g
c) 78.9 g
d) none of the above

Pre-Lab Assignment

Name: ____ **Date:**

1. Why is it important to use separate capillary tubes?

2. Why is it important to keep solution spot with the capillary as small as possible?

1. Why do we cover the chromatography chamber or beaker with a watch glass or aluminum foil?

2. Define Rf.

3. Fill in the blanks:

   a. Color development helps us to locate and identify the movement of cations, so ___________________ reagent is used for the purpose.

   b. DMG stands for____

   c. The paper is placed in a ____________ phase, which is a volatile liquid or a mixture of liquids.

   d. The stationary phase is ___________________.

Name:________________________________   Date:__________________

Objective:

1. To perform chromatographic separation of metal cations in a mixture
2. To identify an unknown mixture of metal ions
3. To learn the difference between physical and chemical changes.

Introduction:

Paper chromatography is a simple, but powerful technique used to separate and identify chemical substances. A mixture to be separated and analyzed is dissolved in a suitable solvent and a single spot is placed on a paper. The paper is placed in a mobile phase, a volatile liquid or a mixture of liquids. The liquid moves upward and in the process, separates the pure components from each other.

Ever wondered why candies are different colors?

Many candies contain colored dyes. Bags of M&Ms or Skittles contain candies of various colors. The labels tell us the names of the dyes used in the candies.

But which dyes are used in which candies?

We can answer this by dissolving the dyes out of the candies and separating them using a method called chromatography. The solvent rises up the chromatography paper (blotting paper) by capillarity. When the solvent reaches the "spot" it dissolves the mixture of colored chemicals. There is now a solution; this is a mixture of solutes dissolved in the solvent. The molecules of these different chemicals are all different sizes. The simple explanation is that the smallest solute molecules travel almost as quickly as the solvent molecules and so get carried to the top of the chromatogram. The largest solute molecules travel very slowly and stay near the bottom, so some of the colored chemical travel further than others. To be objective, retention factor or Rf needs to be defined. It is the ratio of distance traveled by a pure substance to the distance traveled by the solvent used. Each pure substance has unique value of Rf in a suitable solvent. Color development helps us to locate and identify the movement, so 'chromogenic' reagents will be used for this purpose. They will change the color of the cations, but not that of the filter paper.

The retention is a measure of the speed at which a substance moves in a chromatographic system. In continuous development systems like HPLC or GC, where the compounds are eluted with the eluent, the retention is usually measured as the *retention time* R_t or t_R, the time between injection and detection. In interrupted development systems like TLC the retention is measured as the *retention factor* R_f, the run length of the compound divided by the run length of the eluent front:

$$R_f = \frac{distance\ moved\ by\ compound}{distance\ moved\ by\ eluent}$$

The retention of a compound often differs considerably between experiments and laboratories due to variations of the eluent, the stationary phase, temperature, and the setup. It is therefore important to compare the retention of the test compound to that of one or more standard compounds under absolutely identical conditions.

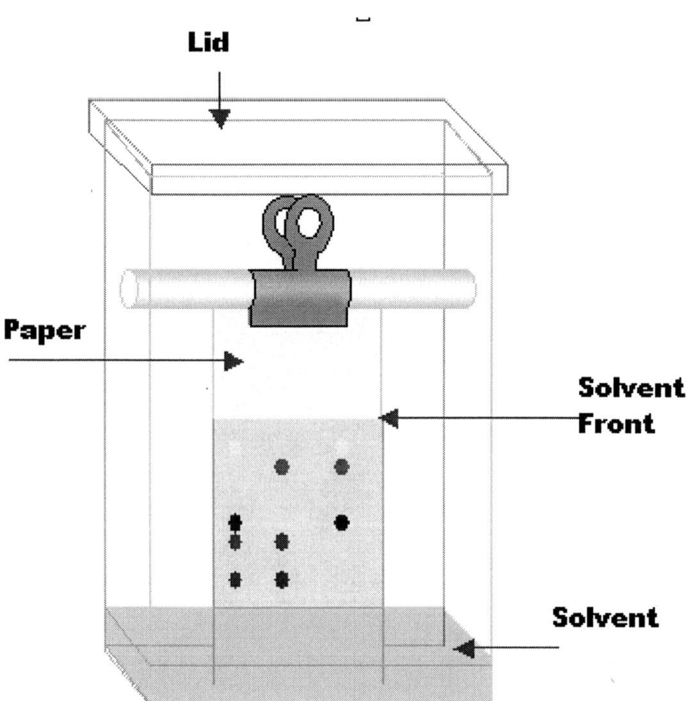

Procedure:

A: Separation of metal cations:

Choose an appropriate size of hand gloves. Take four filter papers and cut each to an appropriate size, so that each can fit in the biggest beaker available to you.

Obtain the samples of known metal cations and an unknown mixture. Record the unknown number. Draw a light pencil line about two cm from both ends of each filter paper. On one line of the filter paper, make five markings at equal distances. These are the positions at which spots will be applied. Under first, second and third spot, write Cu, Ni and Fe with the pencil. On fourth spot write 'un' for unknown mixture and on the fifth spot, write 'st' for standard. You have four filter papers ready now.

Choose four capillary tubes, one for each of the solutions. Carefully, pick up a drop of each metal cation first Cu, then with a new capillary Ni and similarly use the third capillary for Fe. Place each spot on the assigned and marked position on each filter paper. Remember not to contaminate the solutions, so you must use one capillary for only one solution only, but you can use one capillary four times, once for each filter paper. At fourth marked position, (un) carefully spot the unknown mixture, in each filter paper. The last position assigned 'st', mark a little dot with a red, black or blue ball point ink. This ink should serve as signal to stop at the end of the chromatographic separation. Allow all the spots to dry. Bend each filter paper into a cylinder. Do not overlap the edges and staple the ends of the filter paper.

Clean four biggest beakers available as chromatography chambers and add 5 to 10 ml of developing liquid provided to each beaker. This developing liquid is a mixture of hydrochloric acid and acetone. It is flammable and volatile. Place the filter paper cylinder in each beaker so that the metal cation spots are at the bottom side. Cover each beaker either with a watch glass or an aluminum foil. Allow the solvent to rise in the filter paper. As soon as the ink spot reaches the upper pencil line, remove the filter paper and let it dry.

B: Chemical Identification of Metal Ions:

Bring four reagents for use. They are hydrochloric acid (HCl), ammonia (NH_3), potassium hexacyanoferrate (II) (K_4Fe $(CN)_6$) and dimethylglyoxime (DMG).

Use the first filter paper from part-A, and streak or spray it with HCl. Observe any change. Record your observations, immediately, in the experimental data sheet. Let it dry on the side. Take the second filter paper from part-A and expose it to ammonia. Observe any change and record immediately. Let it dry on the side.
Take the third filter paper from part-A and either streak or spray with potassium hexacyanoferrate (II). Observe any change in color on the filter paper and record your observations. Using dimethyl glyoxime, streak or spray the fourth filter paper. Observe any change and record it on the experimental data sheet. As soon as the filter papers from step4 and 5 are dry, expose them to ammonia. What do you see? Record it.
Dry each filter paper (also called chromatogram) with a heat gun and using a pencil, circle the spots with pencils wherever they have moved.

Experiment: 3 Paper Chromatography: Separation Technique

Experimental Data:

Name:_____________   Date:________________

Unknown number: _________

Record all your color observations in the following manner.
a. Did color change?
b. Did color develop?
c. Did color disappear?

____

1. When you sprayed HCl, your observation, record here:

Cu_______________________________
Ni_______________________________
Fe___________________________
Unknown_______________________________

2. When you sprayed NH_3, your observation, record here:

Cu_______________________________
Ni_______________________________
Fe___________________________
Unknown_______________________________

3. When you sprayed $K_4Fe(CN)_6$, your observation, record here:

Cu_______________________________
Ni_______________________________
Fe___________________________
Unknown_______________________________

4. When you sprayed DMG, your observation, record here:

Cu______________________________

Ni______________________________

Fe________________________

Unknown__________________________

5. When you sprayed $K_4Fe(CN)_6 + NH_3$, your observation, record here:

Cu______________________________

Ni______________________________

Fe________________________

Unknown__________________________

6. When you sprayed $DMG + NH_3$, your observation, record here:

Cu______________________________

Ni______________________________

Fe________________________

Unknown__________________________

After recording your observations, answer the following:

1. What is the best reagent to detect the Copper cation?__________________

2. What is the best reagent to detect the Nickel cation?____________________

3. What is the best reagent to detect the Iron cation?______________________________

4. Record the following values:

1. Rf for Copper=__________________
2. Rf for Nickel=____________________
3. Rf for Iron=____________________
4. Rf for unknown=______________________________

2. Circle what you detected in your unknown metal mixture below:

a.   Copper,
b.   Nickel,
c.   Iron
d.   Copper and Nickel
e.   Copper and Iron
f.   Nickel and Iron
g.   Copper, Nickel and Iron

<u>Conclusion:</u>

<u>**Experiment 4 Separation of Mixtures**</u>

<u>**Pre-Lab Assignment**</u>

<u>**Name:**</u>   <u>**Date:**</u>

1 Classify the following as a physical or chemical change?

a. Oxygen gas is cooled to liquid form.
b. Ice evaporates.
c. Grape ferments to produce alcohol.
d. Tums tablets dissolves in water releasing bubbles.
e. Wood log burns in the fireplace.

2 Choose correct word and fill in the blanks:

(Atom of an element or Molecule of a compound,)

Example: Co is Atom of an element, but CO is Molecule of a compound
CO_2 is ______________, and H_2O is a ______________.
K is a ___________ of element Potassium, but NH_3 is a ______________ of ammonia.

3 Classify the following substances either as an element or compound or mixture:

a. Water
b. Silver
c. Table salt
d. Oxygen
e. Air
f. Sugar

4. Classify the following substances either as compound or mixture:

    a. Soil
    b. Air
    c. Water
    d. Wine
    e. Blood

6. Classify either as a pure substance (PS) or homogeneous mixture (HO) or
    heterogeneous (HE) mixture:

a. Pizza
1) PS   2) HO 3) HE

b. Wine
1) PS   2) HO 3) HE

c. Sugar
1) PS 2) HO 3) HE

d. Iodized salt
1) PS 2) HO 3) HE

e. Sodium chloride
1) PS 2) HO 3) HE

7.   The college's name is Lake Michigan. Please choose as many letters from this name,
     but only once and write the symbol and name of the corresponding elements.
     (For example: La is a symbol and it is for element Lanthanum.)

8.   Write your first name and repeat the steps above, to choose as many letters from your
     name, but only once and write the symbol and name of the corresponding elements.

9.   Identify two names of elements from either name of cities, states or famous person.
     Now write the symbol and name of the corresponding elements.

<u>**Experiment 4 Separation of Mixtures**</u>

<u>**Objective:**</u> Learn the difference between pure substance and mixture. Develop skills of separating a homogeneous mixture into pure components and find % recovery.

<u>**Introduction:**</u>

Many substances are pure. Copper, sugar, water and chlorine gas, are examples of pure substances. All elements and compounds are pure substances. Chlorine and copper are elements, while sugar and water represent compounds. Most of the matters we use are not pure substances. Wine and tea, pizza and milk are not pure substances. They are mixtures. A mixture is made up of more than one pure substance.

Some mixtures are homogeneous, like water and salt. They have uniform properties throughout the constituents. Mixtures like pizza and soil are considered as heterogeneous, due to the fact that their composition differs from one part to the other. Milk and oil will make another type of a heterogeneous mixture. Such mixtures are easy to separate.

Most materials used by us and present in our world are mixtures. Very few materials are pure substances. The art of separating mixtures is important because it enables us to isolate pure substances. Mixtures are either homogeneous or heterogeneous. Homogeneous mixtures are uniform in composition. Heterogeneous mixtures are not. Salt water is a mixture of water and NaCl and is homogeneous if thoroughly mixed, with all the salt dissolved. Oil in water is a heterogeneous mixture. Both types of mixtures can be separated into their component parts by physical means.

A salt water mixture can be separated by distilling or evaporating the water and collecting the salt residue. An oil and water mixture will separate into an oil layer and a water layer because the materials are not attracted to one another and gravity "pulls" the denser water beneath the less dense oil. Settling, filtration, chromatography, and manual methods are all means of separating the components of a mixture. Choice of method depends on the type of mixture and the characteristics of its components.

A heterogeneous mixture of solid and liquid or solid and gas is usually fairly easy to separate because of the different phases. The solid may settle out, allowing you to pour off the liquid. If volatile liquid is there, it can be evaporated, leaving the solid behind. Or the mixture can be poured through a filter, catching the solid on the filter and allowing the liquid or gas to pass through. We use filtration frequently in our coffee makers, and cooking.

A mixture of two or more solids is usually separated by utilizing the different chemical or physical properties of the substances. For example, a heterogeneous mixture of corns and white table salt might be separated easily. But what could be done with a mixture of sand and sugar? True, you could get a magnifying glass and tweezers and try picking out the grains of sand, but is there an easier way? Is there some property that sugar has that sand does not (or vice versa)? Could this be used to separate sand and sugar? If you said that sugar dissolves in water and sand does not, you are on the right track.

Homogeneous mixtures of a solvent and one or more solutes (dissolved substances) are often separated by chromatography. Chromatography works to separate a mixture because the components of a mixture distribute themselves differently when they are in contact with a "two phase system". One phase is stationary and the other is moving or mobile. The stationary phase may be a solid packed in a tube or a piece of paper. The mobile phase may be liquid of gaseous. Food colorings are one example, a homogeneous mixture of a solvent and a single dye or combination of selected dyes that produce the desired color. Sublimation is volatilization of a solid. Dry ice (solid carbon dioxide) is probably the most familiar example of a solid that sublimes. But water ice can also be converted directly into water vapor without melting, at low pressure. Snow on mountain peaks disappears without moistening the soil.

The separation of mixtures involves filtration, extraction, sublimation and evaporation. You will learn safe and efficient use of the Bunsen burner. After the separation, the yield of three pure substances is calculated to compute the percentage recovery.

Mixtures are made up of two or more pure substances. During the preparation of a new chemical compound, it is necessary to separate the final desired product from the starting material and other chemical reagents. Such mixture may be homogeneous and require applications of more than one technique.

Procedure: Choose an unknown solid mixture. Record the unknown number in the experimental data report sheet. This mixture contains three white solids in different proportions. The three solids are ammonium chloride, sodium chloride and silicon dioxide and they are in the homogeneous mixture.

Take a clean and dry evaporating dish from your tray. Weigh it and record the mass in 4(a) 1. Transfer all of the unknown mixture in the evaporating dish and reweigh. Record the combined mass in 4(a) 2. The difference between two measured masses should be the mass of your unknown mixture. Record it in 4(a) 3.

Take the evaporating dish containing the unknown sample of mixtures to the fume hood. Set up the evaporating dish for heating using a stand, ring clamp and a clay triangle. Check the rubber tubing of the Bunsen burner and if it is good, join it to a gas outlet. Turn the gas outlet to open position and immediately, using the striker, light the burner. It should give a pale blue flame. If it does not, adjust the vent at the bottom of the Bunsen burner. Still if you have problem, consult the instructor. Slowly and carefully, heat the bottom of the evaporating dish. Within few minutes, you should see white milky smoke. This is the process of sublimation, in which a solid passes directly to the gaseous state without the appearance of the liquid state. Continue heating and once white milky smoke stops, it is almost completion of sublimation. Continue careful heating for one more minute, then turn off the Bunsen burner. Cool the evaporating dish to room temperature. Take it to the balance and using the same balance, weigh the remaining solid and the evaporating dish. Record it in 4(b) 1. The difference between 4 (a)2 and 4(b) 1 is the mass of ammonium chloride. Compute and record it 4(b) 2.

Isolation of silicon dioxide using filtration:

Take a filter paper and weigh it. Record the mass in 4(C) 1. Using the distill water from the wash bottle, transfer the remaining mixture from the evaporating dish to a beaker. Once the remaining solid mixture is transferred, using a stirring rod, stir the content in the beaker for three minutes. St up plastic funnel on a metal ring clamp. Fold a filter paper and place it in the plastic funnel. Moisten the filter paper slightly. Place a clean evaporating dish under the funnel. Carry out filtration. If necessary, wash the residue left in the beaker using the distilled water from the wash bottle. Once the filtration is over, carefully remove the wet filter paper, containing solid silicon dioxide, and place it on a dry watch glass. Leave the evaporating dish undisturbed for the last part of isolation of sodium chloride using evaporation.

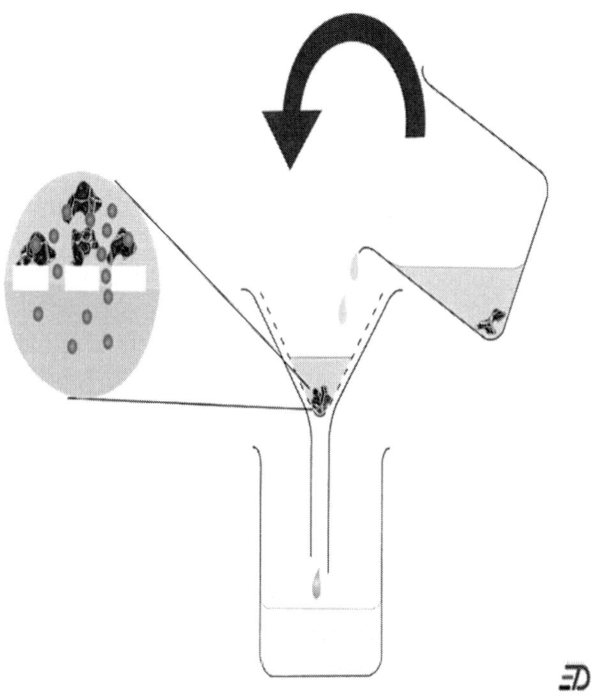

Keep the watch glass with wet filter paper in the microwave oven. Turn on the microwave for few seconds. If the filter paper is not dry, continue heating in microwave oven, till it is dry. You must be standing near and watching it. Do not burn the filter paper. If it remains slightly wet, you can air dry for few minutes. Once the filter paper and silicon dioxide are dry, weigh them. Record the mass in 4 (C) 2. The difference is the mass of silicon dioxide. Compute and record it in 4 (c) 3.
P28

Isolation of sodium chloride using evaporation technique:

Place the evaporating dish containing sodium chloride and distilled water on a clay triangle, over a ring stand. Turn on the Bunsen burner and very gently with due care, start heating the evaporating dish. If it is too much heat, hold the burner in such a way, so as to control the heat.

Heat till majority of water has evaporated. Turn off the burner and using crucible tong, remove the evaporating dish. Take it to the microwave oven and heat for few seconds. If necessary, repeat till sodium chloride in the evaporating dish is completely dry. Remove from the microwave oven and let it cool to room temperature. Weigh the evaporating dish with sodium chloride. Record the mass in 4 (d) 1. The difference between 4 (d) 1 and 4 (a) 1 is the mass of sodium chloride. Calculate and record it in 4 (d) 2.

Applying the formula shown below, compute the % of each pure substance isolated and record the results.

$$\% \text{ of pure substance} = \frac{\text{Mass of pure substance}}{\text{Mass of the mixture in sample}} \times 100$$

Dispose all the three pure substances isolated from the mixture as per instructor's directions.

<u>**Experiment 4 Separation of Mixtures**</u>
<u>**Experimental Data:**</u>

Name_______________________________________Date:_______________________________

Unknown number of the sample mixture   ___________

A: Sample mixture:

4 (a) 1. Weight of evaporating dish =   __________ g.
4 (a) 2. Weight of evaporating dish + sample mixture =   __________ g
4 (a) 3. Weight of sample mixture =   __________ g.

B: Ammonium chloride:

4 (b) 1. Weight of evaporating dish -ammonium chloride   =__________ g
(after removing white smoke)
4 (b) 2 Weight of ammonium chloride = {4 (a) 2—4 (b) 1}=__________ g

C. Silicon dioxide:

4 (c) 1 Weight of the filter paper =   __________ g
4 (c) 2 Weight of filter paper + silicon dioxide =   __________ g
4 (C) 3 Weight of silicon dioxide = { 4 (C) 2---4 (c) 1}   __________ g

D. Sodium chloride:

4(D) 1. Weight of evaporating dish + sodium chloride =   __________ g
4 (D) 2 Weight of sodium chloride = { 4 (D) 1 ---4 (a) 1}=__________ g.

Percentage recovery of pure substances =

% of ammonium chloride =   _______________
% of silicon dioxide =   _______________
% of sodium chloride =   _______________

<u>**Question:**</u>
1. Did you recover 100%? If not, state the source of %error in this experiment.

2. If an unknown sample contains 1.50 g mixture. What is the percentage recovery of matter, if a student separates the three pure substances as shown below;

a) Amount of ammonium chloride = 0.28 g
b) Amount of silicon dioxide = 0.91 g
c) Amount of sodium chloride = 0.20 g

3. Explain the difference between a pure substance and a mixture. From the following mixtures and identify pure substances in each one of them.

a) Iodized table salt

Pure substances:__________________ and ______________

b) Air

Pure substances:________________, ________________and ____________

4. Explain about other separation techniques used to purify solids and liquid mixtures. You can draw and label them.

<u>**Experiment #5**</u>   **Temperature Changes: Heat Capacity**

Name:_________________________   Date:_____________________________

<u>**Pre-Lab Assignment:**</u>

1. What are the basic components of a calorimeter experiment?

2. A 30.0 g sample of a metal was heated in a hot water bath to 80°C. It was then quickly transferred to a coffee-cup calorimeter. The calorimeter contained 100.0 g of water at a temperature of 20°C. The final temperature of the contents of the calorimeter was 25°C. What is the specific heat capacity of the metal?

a)   -0.66 J/g·°C
b)   1.27 J/g·°C
c)   -1.27 J/g·°C
d)   -11.4 J/g·°C

3. Look at Table 3.4 in your textbook "Specific Heat Capacities" and answer the following questions:

a) Which substance has the highest specific heat capacity?___________________

b) Which substance has the lowest specific heat capacity?___________________

4. Look at Table 3.4 in your textbook "Specific Heat Capacities" and answer. If you have 100 L water and 100 L ethanol and you want to boil them for an experiment. Which liquid will be easy and faster to boil?

OBJECTIVES

1. Learn to construct and use a simple calorimeter. •
2. Learn to measure temperature using a temperature probe.
3. Determine heat lost by cooling water and heat gained by warming water and learn to relate energy, temperature change, and heat capacity.

Introduction:

Heat is energy transferred between matters. The transfer occurs because of differences in temperature. The ability of matter to transfer heat depends on three factors:
1. Mass
2. Temperature
3. Composition.
To study heat transfer chemists use a calorimeter. It is an instrument used to measure changes in heat energy. You can make a simple calorimeter using a Styrofoam cup. In this experiment, we will use a Temperature Probe to measure changes in temperatures. The joule (J) is the SI unit for heat energy, just like calories in English system. To calculate change in heat energy, we can use:

Heat= Mass x Specific Heat Capacity x Temperature Change

Let us take one example to understand energy, temperature change, and heat capacity. Say, you want to find out how much energy in the form of heat is needed to warm 235 g of water from room temperature of 25^0 C to boiling temperature of 100^0 C.

We can apply,

Heat= Mass x Specific Heat Capacity x Temperature Change

Mass= 235 grams
Specific Heat Capacity for water= 4.18 J/g ^{0}C
Temperature Change=100 –25= 75 ^{0}C

Heat= (235 g) x (4.18J/ g ^{0}C) x (75 ^{0}C) = 7.4 x 10^4 J

Note that each number had units and as we calculated energy value, we cancel equal
units.

<u>Procedure:</u>

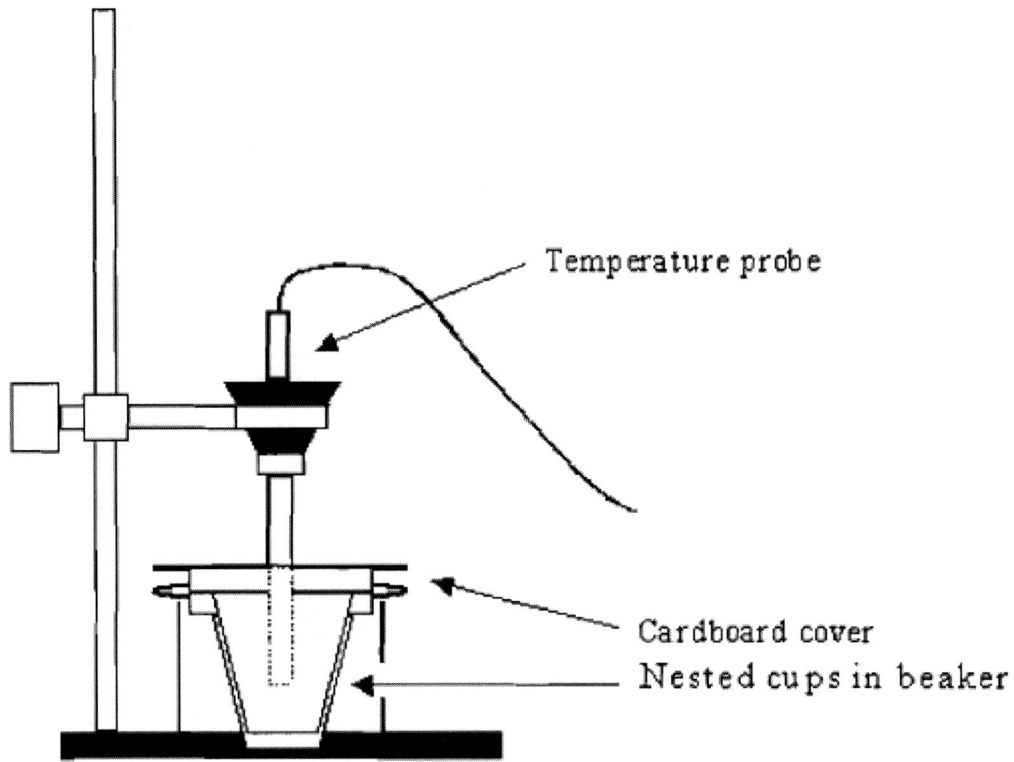

1. Obtain and connect the probes to the computer interface. Prepare the computer for data
collection by opening the file "07 Mixing Warm Cold".
2. Place a Styrofoam cup into a 250 mL beaker.
3. Now using a graduated cylinder, get 50.0 mL (50.0 g) of ice cold water Pour the cold
water into the Styrofoam cup and insert Probe 1.
4. Use another graduated cylinder to get 50.0 mL (50.0 g) of warm water and place Probe
2 into the warm water in that graduated cylinder for one minute. After the probes have
been in the cold and warm water, begin data collection.
5. Now we are ready to measure the temperature using the computer.
6. Click to begin data collection. After the first temperature readings, transfer the warm
water and its probe to the Styrofoam cup. Stir to mix the warm water with the cold water,
and then continue to collect data until both temperatures stop changing.
7. Click the Statistics button.
8. Record the minimum and maximum temperatures.

<u>Experimental Data:</u>

1. Cold Water Probe 1 MIN _____________ °C MAX __________ °C
2. Warm Water Probe 2 MIN __________ °C MAX __________ °C
3. Mixing _________________°C

4. The temperature change, Δt, ___________________

5. The heat gained by the cold water (in J). _______________

Use the equation $Q = \Delta t \cdot m \cdot \text{Specific Heat Capacity}$

Where Q = heat absorbed (in J),
Δt = change in temperature (in °C),
m = mass (50.0 g in this experiment), and
Specific heat capacity is (4.18 J/g°C for water).

6. The heat lost by the warm water_______________ (in J).

7. % difference:

Calculate the percent difference using the formula
% difference = (heat lost – heat gained)/heat loss (100)

Questions:

1. What is the law of conservation of energy?

2. Heat is thermal energy, give examples of different forms of energy.

Lab 6 Mole Calculation and Determination of a Hydrate Formula

Pre-Lab Assignment

Name: _______________________   Date:

1. A hydrate with formula $CuI .n\ H_2O$ weighs 160.20 grams before heating. On heating, water was completely lost and the weight of anhydrous salt CuI was found to be 90.30 grams. Calculate the value of n in $CuI .n\ H_2O$.

2. Calculate the number of grams in 1.5 moles of $MgSO_4$

3. How many milligrams are present in 2.35 moles of water?

4. One student heated 20.0 grams of the hydrate, instead of 2.0 grams. He worked for six hours in the lab, but his results came out accurate. How that is possible?

1. A hydrate weighing 6.02 g of $CuSO_4.n\ H_2O$ is heated in a crucible. After cooling to room temperature, the mass was found to be only 3.09 g.
   a. Determine the mass of water driven from the hydrate during the heating process.
   b. Determine the mass of anhydrous salt left in the crucible after the heating process.
   c. Determine the percentage of water in the hydrate.
   d. Determine the empirical formula of the hydrated salt.

Lab 6. Mole Calculation and Determination of a Hydrate Formula

Objective:

First by calculating moles of water and then determining the formula of a hydrate, we understand the concept of mole.

Introduction:

Donuts and eggs are sold by dozens. One dozen is equal to 12. Chemists use the term mole to represent a number. It is 6.02×10^{23}. This number is also called Avagadro's number. This means that one mole of any element consists of 6.02×10^{23} atoms of that element. A dozen of anything contains 12 items, similarly a mole has 6.02×10^{23} particles. One mole is also equal to the molar mass of a compound. So one mole of water has a mass of 18 grams and it contains 6.02×10^{23} molecules of water.

CuSO4 nH2O (s)   + Heat →   CuSO4 (s)   + n H2O(g)

What is the mass of 5.0 moles of NH₃ (Ammonia)?

Solution:

First of all, we need to know the mass of grams of one mole of ammonia (NH_3), as we calculated for water and glucose.

Atomic weight of N + (Atomic weight of H) x 3
=   14 +   (1) x 3
=   14 +   3
=   17

Note that 1.0 mole of ammonia has the mass of 17 g,
So;   5.0 mole of ammonia will weigh
 17 g x 5 mole/ 1 mole
= 85 g.

Determine the number of moles of 368 grams of H_2SO_4.

<u>Solution:</u>

First, look up the atomic masses (relative atomic mass) for hydrogen, sulfur and oxygen from the periodic table.

Atomic mass of hydrogen = 1.01 g/mole
Atomic mass of sulfur = 32.07 g/mole
Atomic mass of oxygen = 16.00 g/mole

The formula mass of H_2SO_4 = 1.01*2 + 32.07 + 16.00 *4 = 98.09 g/mole

1 mole of H2SO4 = weights 98.09 grams.
Moles H_2SO_4 (368g) Calculation: (368g)/ (98.09g/mole) = 3.75 moles of H_2SO_4

Using the relationship between the numbers of particles present in one mole, we are able to find out exact number of molecules and atoms present in given number of moles.

Procedure:

1. Clean a crucible. Dry it. Arrange it on a clay triangle on a metal ring. Turn on the Bunsen burner and arrange it under the ring. Heat the crucible for two to five minutes. After five minutes, remove heat and let the crucible cool. Once it is cool, weigh it. Record the mass in experimental data (1).

2. From the community table, take some solid hydrate in a clean beaker or evaporating dish. Add approximately 2 grams of the hydrate to the crucible. Record the mass of the hydrate in (2).

3. Set up the crucible containing the hydrate on the clay triangle and partially cover it with a lid. Heat slowly, for 2 minutes then increase heat moderately. Heat for about five minutes and then remove heat. Cover the crucible completely with the lid and let it cool to room temperature. Weigh the crucible and content. Record the mass of the anhydrous solid remaining in the crucible. (3).

4. Again set up the crucible, anhydrous solid on the clay triangle on the ring clamp. Reheat the crucible leaving the lid little open. Turn off the heat after about 2 minutes of heating. Let the crucible cool to room temperature and then reweigh it. Record the mass (4).

NOTE: If the two readings 9.3 and 9.4 are within 5 % difference, conclude the experiment. If the two readings do not agree, repeat the step (5).

Lab 6. Mole Calculation and Determination of a Hydrate Formula

Name:_________________________________   Date:______________

Experimental Data:

1. Weight of clean crucible+ cover: ______________________g

2. Weight of crucible + cover + hydrate   ______________ g

3. Weight of hydrate (#2 --- #1)   _______ _________g

4. Weight of crucible + anhydrous salt (after heating for 5 minutes) + cover ________g

5. Weight of water lost: (#2 -- #4)   ______________ g.

6. Weight of crucible + anhydrous salt (after second heating for 2 minutes)

          ________g

7. Weight of water lost: (#2--#6)   ______________ g.

8. Average weight of water lost: (Numbers from #5+#7 divided by 2)
________________g

9. Weight of anhydrous salt (#3--#8) _______________ g.

10. Moles of the water :(#8 divided by 18)=_________________

11. Moles of hydrate: (#9 divided by 159.6)=_________________

Now calculate formula:

#11 divided by #11 = 1
#10 divide by # 11= n

Write value of n in the following blank

Formula is: $CuSO_4$ ______H_2O

Experiment: 7 Series of Chemical Reactions

Name:_____________________________   **Date:**_________________________________

Pre-Lab Assignment:

1. Balance the following equations:

a) C_2H_6 + O_2 → CO_2 + H_2O
b)   As   + O_2 → As_2O_5
c) H_2 + Cl_2 →   HCl
d)   KI   + Br_2 →   KBr   + I_2

2. Balance and classify chemical reactions:
(Remember that reactions may belong to more than one category.)

$KClO_3$ (s) → KCl (s) + O_2 (g)_____________________________

Fe + O_2 → Fe_2O_3____________________________________

HBr +   KOH →   KBr + H_2O____________________________

Zn (s) + $Al(NO_3)_3$ → $Zn(NO_3)_2$ +   Al (s)______________________________

3. A reaction (Remember that reactions can be belong to more than one category.)

$Pb(NO_3)_2$ (aq) → PbO_2 + 2 NO_2 (g) can be classified as

____.

4. Write total and net ionic equation for a reaction

NaCl (aq)   + $AgNO_3$ (aq)   → $NaNO_3$ (aq)   +   AgCl(s)

a. Total:

b. Net:

5. Balance: Al   +   O$_2$ →   Al$_2$O$_3$

____

6. To balance an equation like

C$_8$H$_{18}$ +   O$_2$ → CO$_2$ +   H$_2$O

You need a coefficient in front like
___________ C$_8$H$_{18}$
___________ O$_2$
___________CO$_2$

7. When an acid combines with a base, the reaction is called___________________________

8. Balance with appropriate coefficients:

1. __NaCl +   __BeF$_2$ →   __NaF +   __BeCl$_2$
2. __FeCl$_3$ +   __Be$_3$(PO$_4$)$_2$ →   __BeCl$_2$ +   __FePO$_4$
3. __AgNO$_3$ +   __LiOH →   __AgOH +   __LiNO$_3$
4. __CH$_4$ +   __O$_2$ →   __CO$_2$ +   __H$_2$O
5. __Mg +   __Mn$_2$O$_3$ →   _MgO +   __Mn

9. Show the following equation in molecular equation, total ionic equation and net ionic equation presentations:

"Aqueous potassium phosphate reacts with aqueous silver nitrate to produce aqueous potassium nitrate and precipitate of silver phosphate."

a. molecular equation, ____

b. total ionic equation ____

c. net ionic equation____

10. Balance:

___________Al$_{(s)}$ +_________CuCl$_2$ $_{(aq)}$ →__________AlCl$_3$ (aq) +_______Cu$_{(s)}$

11. Identify the following chemical equations either as balanced ("B) or unbalanced ("U").

B,   U   a. N_2 + H_2 → $2NH_3$

B,   U   b. $2\,Mg$ + O_2 → MgO

B,   U   d. $2C$ + 0_2 → $2CO_2$

12. Identify the following chemical equations, as Combination (C) or Decomposition (D):

C,   D   1. $2HgO(s)$ → $2Hg(l)$ + $O_2\ (g)$

C,   D   2. $2\,KBrO_3$ → $2KBr$ + $3O_2$

C,   D   3. Cd + S → $CdS(s)$

13. Identify the following equations as single replacement (S) or double replacement (D):

S,   D   1. $CuSO_{4(aq)}$ + $Zn\ (s)$ → $ZnSO_4\ (aq)$ + $CU(s)$

S,   D   2. Fe_2O_3 + $3\,C$ → Fe + $3\,CO$

S,   D   3. $AgNO(aq)$ + $NaCl(aq)$ → $AgCl(s)$ + $NaNO_{3(aq)}$

14. Classify chemical reactions:

      1) $2\,KClO_3\ (s)$ → $2\,KCl\ (s)$ + $3\,O_2\ (g)$_______________

      2) $4\,Fe + 3\,O_2$ → $2\,Fe_2O_3$_______________

      3) HBr + KOH → KBr + H_2O_______________

      4) $Zn\ (s) + Al(NO_3)_3$ → $Zn(NO_3)_2$ + $Al\ (s)$_______________

      5) $Cu + O_2$ → Cu_2O_______________

15. Balance each of the equations from the formulas given:

    a) $CaCl_2 + AgNO_3$ ----> $AgCl + Ca(NO_3)_2$

    b) $H_2 + O_2$ ----> H_2O

    c) $Mg + P_4$ ---> Mg_3P_2

    d) $H_2SO_4 + NaOH$ ----> $H_2O + Na_2SO_4$

    e) $Zn + HCl$ ---> $ZnCl_2 + H_2$

    f) $KClO_3$ ---> $KCl + O_2$

    g) $Fe + O_2$ ---> Fe_2O_3

    h) $C_2H_6 + O_2$ ---> $CO_2 + H_2O$

16. For each of the following reactions, determine what the products of each reaction will be.

1)　　_____ $Ca(OH)_2$ + _____ HF → ___________________________

2)　　_____ $Pb(NO_3)_2$ + _____ K_2CrO_4 → ___________________________

3)　　_____ NaF + _____ H_2SO_4 → ___________________________

4)　　_____ $Cu(OH)_2$ + _____ H_3PO_4 → ___________________________

5)　　_____ $AgNO_3$ + _____ Na_2CO_3 → ___________________________

6)　　_____ Zn + _____ H_2CO_3 → ___________________________

7)　　_____ $Pb(OH)_2$ + _____ Hg_2S → ___________________________

<u>17. Balance the equations below:</u>

1) _____ N_2 + _____ H_2 → _____ NH_3

2) _____ $KClO_3$ → _____ KCl + _____ O_2

3) _____ $NaCl$ + _____ F_2 → _____ NaF + _____ Cl_2

4) _____ H_2 + _____ O_2 → _____ H_2O

5) _____ $Pb(OH)_2$ + _____ HCl → _____ H_2O + _____ $PbCl_2$

6) _____ $AlBr_3$ + _____ K_2SO_4 → _____ KBr + _____ $Al_2(SO_4)_3$

7) _____ CH_4 + _____ O_2 → _____ CO_2 + _____ H_2O

8) _____ C_3H_8 + _____ O_2 → _____ CO_2 + _____ H_2O

9) _____ C_8H_{18} + _____ O_2 → _____ CO_2 + _____ H_2O

10) _____ $FeCl_3$ + _____ $NaOH$ → _____ $Fe(OH)_3$ + _____$NaCl$

Experiment7 Series of Chemical Reactions

Objective:

To learn the skill to carry out a chemical reaction, describe it by writing molecular, total ionic and net ionic equations and be able to classify it.

Introduction:
Chemical changes are called chemical reactions. Instead of describing the chemical reactions, in words, chemists have devised a more efficient method, symbols and formula to represent elements and compounds. To show a chemical change, when someone uses symbols of elements and formula of compounds, it is a chemical equation.

The following guidelines will help us to balance the chemical equations:

1. The correct symbol and formula must be written.
2. The subscript in the formula should not be changed.
3. First of all, try to balance the element with the highest number of atoms.
4. Use the lowest possible balancing coefficient and write in front of the symbol or formula.
5. Balance the diatomic gas last.

Let us write an equation and balance it. Methane gas (CH_4) reacts with O_2 (g) to produce Carbon dioxide gas and water.

$$CH_4 \,(g) + O_2 \,(g) \;\rightarrow\; CO_2 \,(g) + H_2O \,(l)$$

Methane and oxygen are the two reactants, while carbon dioxide and water are the products. Note the subscript '4' for hydrogen in methane. This is the highest number of atoms in the three elements, carbon, hydrogen and oxygen, present. There are only '2' hydrogen atoms in water on the right side of the arrow. If we use a balancing coefficient of '2' and write in the front of formula of water, that will balance hydrogen atoms on the both sides of the arrow. Let us do it, now.

$$CH_4 \,(g) + O_2 \,(g) \rightarrow CO_2(g) + 2\,H_2O(l)$$

Now the next element, which should be balanced, is carbon. But it is already balanced. It means the diatomic gas, oxygen, has to be balanced the last. There are 2 atoms of oxygen on the left side of the arrow and four atoms on the product side. If we write 2 in front of oxygen on the reactant side, the number of atoms of oxygen is balanced on both sides.

$$CH_4\,(g) + 2O_2\,(g) \rightarrow CO_2\,(g) + 2H_2O(l)$$

1 carbon, 4 "H" and 4 "O" atoms = 1 carbon, 2+2 (total 4) "O" and 4 "H" atoms. Inspecting the equation and counting the number of atoms present, we find the following facts: There is one carbon atom on each side. On both sides of the arrow, numbers of hydrogen atoms and oxygen atoms are four. And now it is a balanced equation.

Procedure:

<u>Copper (II) Nitrate to Copper (II) Hydroxide</u>

Place the test tube in a beaker of ice water. Add 2 mL of 6 M sodium hydroxide. Stir with a glass stir rod. To test whether enough sodium hydroxide has been added to react with all of the copper, touch the stir rod on a piece of pink litmus paper. If the litmus turns blue, enough sodium hydroxide has been added. If the litmus remains pink, add more sodium hydroxide and repeat the litmus test. The blue precipitate that forms is copper (II) hydroxide.

Write the balanced chemical equation, total ionic and net ionic equation of Copper (II) Nitrate + Sodium Hydroxide and record in Table 1.

<u>Copper (II) Hydroxide to Copper (II) Oxide</u>

Using a Bunsen burner, heat water in a beaker, once it starts boiling, remove burner and stop heating , Place the test tube in the boiling water and stir the copper (II) hydroxide until all of the copper (II) hydroxide turns to black copper (II) oxide (CuO).

Place the test tube in a centrifuge. Balance the centrifuge by placing a test tube with an approximately equivalent amount of water opposite your test tube in the centrifuge. Centrifuge the test tube and carefully remove the liquid with a plastic dropper. Fill the test tube ¾ full with distilled water, stir and centrifuge the test tube. Again remove the liquid with the dropper.

<u>Copper (II) Oxide to Copper (II) Sulfate</u>

Add about 3 mL 3 M sulfuric acid (H_2SO_4) to the tube and stir the contents. The black copper (II) oxide will dissolve to give a blue solution of copper (II) sulfate.

<u>Copper (II) Sulfate to Copper</u>
Slowly add a small amount of granular zinc metal to the copper solution. Occasionally
stir with the stir rod. Spongy red copper will deposit in the bottom of the tube. The blue
coloration will disappear as the copper ions change to copper atoms.

As the copper forms, you will also observe H_2 bubbles being produced as a result of the
reaction between zinc and the sulfuric acid. Leave it in the fume hood.

Lab Experiment: 7 Series of Chemical Reactions

Name:_______________________________   Date:_____________________________________

<u>Lab Report:</u>

<u>Results and Conclusion:</u>

a) Write the balanced chemical equation, total ionic and net ionic equation for each of the first reaction

b) Record your observations for each of the following reactions
Did color change?
Did solid form?
Did any gas or bubbles form?

<u>Table 1Copper (II) Nitrate to Copper (II) Hydroxide</u>

<u>Observations   Type of Reaction</u>

<u>1.</u>

<u>2.</u>

<u>3.</u>

2. <u>Copper (II) Hydroxide to Copper (II) Oxide</u>

a. Write the balanced chemical equation,
b. Record observations.

3. Copper (II) Oxide to Copper (II) Sulfate

Write the balanced chemical equation and Record observation.

---

4. Copper (II) Sulfate to Copper

Write the balanced chemical equation and Record observation.

.

---

5. Zinc + Sulfuric Acid

Write the balanced chemical equation and record observation.

Questions:

1. Why is this series of reactions often called the "copper cycle"?

2. Copper recovered is recycled for next year's lab. Think about all types of materials that you know are recycled. List them.

3. What types of materials are too expensive to recycle?

4. What is the purpose of using centrifuge? Can it be used to separate different isotopes of an element? Which isotope will be deposited at the bottom of the tube, lighter or heavier?

Experiment 8 Atomic Structure and Flame Spectra
Name:_____________________ **Date:**___________________

Pre-Lab Assignment:

1. Which is the smallest alkali metal? Which is the smallest halogen element?

2. Write the electronic configuration of the alkali metal with atomic number 11 and halogen with atomic number 17.

3. Which atom(s) has unpaired electron(s) among, carbon, nitrogen, oxygen and fluorine? How many unpaired electron(s)? List all possible answers.

4. a) You are cooking soup with table salt and boils over on the stove. It burns with a bright golden yellow flame, what must be the reason?

b) Compare atom with a city and show similarity in them.

5. Prepare a graph by plotting the atomic number of the first ten elements on the horizontal axis and their corresponding atomic masses on the vertical axis. Using a ruler, connect the points you have on the plot.

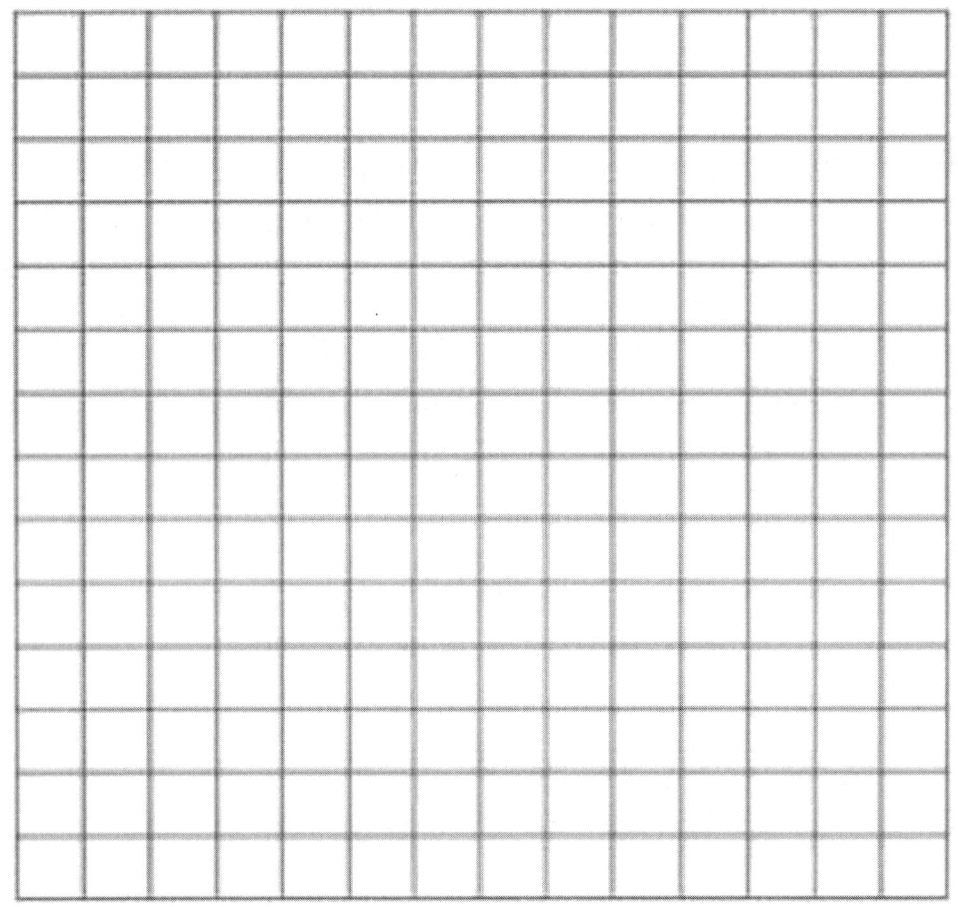

Experiment 8 Atomic Structure and Flame Spectra

Objective:

1. To learn correct symbols and names of the elements and classify them as metals or non-metals.

2. Learn about writing correct electron configuration and understand the concept of isotopes.

3. Using flame spectroscopy, interpret the energy levels in terms of electronic arrangement.

Introduction:

It is important to learn about properties of the elements. The periodic table is an excellent tool, which is organized, in an efficient table. Elements are made up atoms, which in turn consist of subatomic particles. Protons, neutrons and electrons are positive, neutral and negative charged particles. Model of the atom can be simplified in terms of electron arrangement. The electrons from the atom, when excited by energy provided from a flame move up to a higher energy state. Then these electrons fall back giving off energy in the form of light of a characteristic color from blue to red light.

Electron Configuration:

Niels Bohr proposed the arrangement of electrons in atoms of elements. His idea was based on the solar system. Like the planets move around the sun, Bohr explained that electrons move around the nucleus of the atom, in circular orbits. Research revealed that Bohr's model worked for simple atom like hydrogen, but a revised atomic model was needed to explain the electron configuration.

Compare the two models of a hydrogen atom shown above. Erwin Schrödinger refined the Bohr model of atom and proposed the quantum mechanical model. Solving wave equations, Schrödinger was able to plot the electron probability distribution of an electron. The probability of locating electrons is highest around the nucleus of an atom.

It is important to remember that another name for orbital is 'sub shells'. The name of these sub shells or orbital come from the words sharp (s), principal (p), and diffuse (d) and fine (f). The shapes of orbital also differ from each other as shown in the diagram. The 's' orbital is spherical, while 'p' orbital is dumbbell shape. Orientation of 'p' orbital can differ in the three-dimensional space. As shown below, when 'p' orbital is along 'x'-axis, it is called P_x and when it is oriented along 'y' and 'z' axes, they are referred to P_y and P_z. The volume or space shown by these orbital represents the location of electrons.

We can study the filling of electrons in orbital by understanding the following simple rules:

1. Electrons fill the lowest energy level orbital first.
2. A maximum of two electrons is allowed in each orbital. Both electrons must have opposite spin this is Pauli's exclusion principle.
3. As long as orbitals of the same energy are available, electrons do not pair-up, this is Hund's rule.

Starting with hydrogen atom, which has only one electron, will occupy 1s orbital.

Hydrogen (H) $1s^1$

Second element 'He' has 2 electrons. Both these electrons will occupy the same orbital, 1s. But both these electrons must have opposite spin.

Helium (He) $1s^2$

<u>Lithium to Neon:</u> The third element, lithium, has atomic number 3. First two electrons of lithium will go in 1s orbital and it will be filled. The third element has to occupy 2s orbital. The electron arrangement of 'Li' will be

Lithium (Li) $1s^2, 2s^1$

Beryllium is the fourth element in the periodic table. It has four electrons. First two electrons will pair-up and occupy 1s orbital and remaining two electrons will also pair-up. They occupy 2s orbital.

The electron configuration of 'Be' will be

Beryllium (Be) $1s^2, 2s^2$

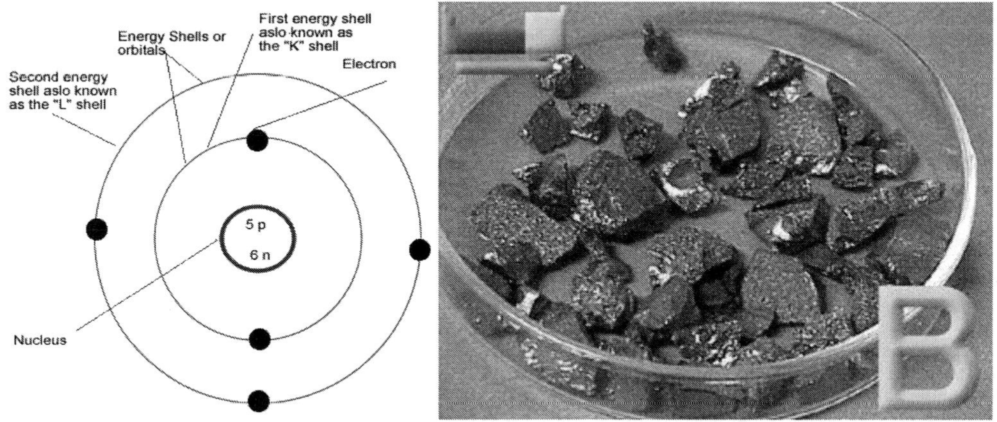

Boron has five electrons. First four electrons will be arranged just like 'Be', but the fifth electron will occupy 2p orbital, so the configuration of 'B' will look like:

Boron (B) $1s^2, 2s^2, 2p^1$

Carbon has atomic number 6. It means there are six electrons. The first four electrons will be paired and placed in 1s and 2s orbital. The last two electrons will not pair-up. According to Hund's rule, one electron will go in $2p_x$ and another will occupy $2p_y$. The electron configuration of carbon will be:

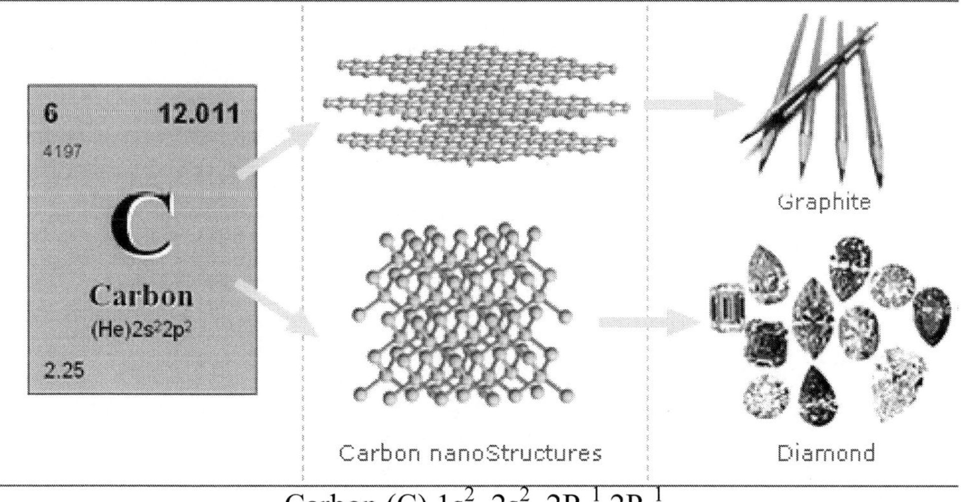

Carbon (C) $1s^2, 2s^2, 2P_x^1, 2P_y^1$

Nitrogen has 7 electrons and again the first four electrons will be arranged just like Be. Remaining three electrons will occupy $2p_x$, $2p_y$, and $2p_z$. The last three electrons will be unpaired. Nitrogen (N) $1s^2, 2s^2, 2P_x^1, 2P_y^1, 2P_z^1$

The next element is oxygen. It has 8 electrons. The first two electrons will be paired in 1s orbital; next two electrons will be paired in 2s orbital. Now out of the four remaining electrons, three will go in $2P_x$, $2P_y$ and $2P_z$. All three electrons will have parallel spin. The last electron has no choice but to pair-up with the electron which is already there in $2P_x$ orbital.

Thus oxygen will have the following electron configuration:

Oxygen (O) $1s^2\, 2s^2,\, 2P_x{}^2,\, 2P_y{}^1,\, 2P_z{}^1$

Fluorine with 9 electrons will have the following configuration:

Fluorine (F) $1s^2,\, 2s^2,\, 2P_x{}^2,\, 2P_y{}^2,\, 2P_z$

Neon has 10 electrons. The first two electrons will complete the first shell and remaining eight electrons will fill-up the second shell. The complete electron arrangement of 'Ne' will be as shown below:

Neon (Ne) $1s^2,\, 2s^2,\, 2P_x{}^2,\, 2P_y{}^2,\, 2P_z{}^2$

The arrangement of electrons explains the periodicity of elements. The similarity of elements of one group in the periodic table is directly linked to the configuration of electrons, particularly in the outermost shell of the atom.

Procedure:

Part-A:

a. Observe the six samples of elements and fill out the information under Table 1.
b. In Table 2, fill out the electron configuration of the elements listed in Table 1.

Part-B:

1. Obtain a spot plate, platinum loop wire and get 15.0 ml of 6.0 M HCl in a 50 ml beaker. Be careful with 6.0 M HCl as it is corrosive.
2. Turn on the Bunsen burner and adjust the flame till it has a blue cone.
3. Dip the platinum wire in 6.0 M HCl and keep it in the blue cone of the flame for two minutes.
4. Repeat cleaning the platinum wire two to three times.
5. When the loop glows red with no sign of contaminating color in the flame, it is clean and ready for the use.
6. Take known and unknown solutions in the spot plate and label them with a marker pen. Record the unknown number.
7. Dip the clean platinum wire loop in the first known solution and keep the loop of the wire into the blue cone of the flame.
8. Observe the color of the flame and if necessary repeat the procedure. Record your observations.
9. Repeat the flame test for all other known solutions as well as the unknown solution. Record observations in Table 2.

10. Conclude the identity of your unknown. Recheck the flame color of unknown and suspected known solution, side by side, with two platinum wires in the flame to confirm the identity. Record it.
11. You can discard all salt solutions in the sink; it is safe to do so.
12. Return all the items from where you got it.

Experiment 8 Atomic Structure and Flame Spectra

Experimental Data:

Name:___________________________Date:________________

Table 1: Fill the missing items:

Physical properties and classification of Elements:

Name	Symbol	Subatomic			
		Proton	Electron	Neutron	
Example:					
Lithium	Li	3	3	4	Metal
Potassium					
Sodium					
Barium					
	Ca				
				38	

Table 2   Electron configuration:

1. K
2.Na
3.Li
4.Mg

<u>**Results:**</u>

My unknown #_______________

The identity of unknown is   cation,

because____

____

____

____

Questions:

1. For the following electron configurations choose possible element (or ions) they may represent

1. $1s^2 2s^2 2p^6 3s^2 3p^6 4s^2 3d^{10} 4p^4$ _______________________________

2. $1s^2 2s^2 2p^6 3s^2 3p^6 4s^2 3d^{10} 4p^5$ _______________________________

3. $1s^2 2s^2 2p^6 3s^2 3p^6 4s^2 3d^{10}$

$4p^6$ ____

2. Note: Valence electrons are ALL the electrons in the last energy level, including s and p electrons. Atoms will either lose all valence electrons, if they have less than four, or they will gain electrons, if they have more than four, to achieve the full and stable electron configuration like the noble gases.

a. Show the complete or long hand electron configuration for fluorine, F.

____

b. How many valence electrons does fluorine have?_______________________

c. Will fluorine tend to gain or lose electrons? Why?______________________

d. What is the charge on the stable fluorine ion? Show the fluorine ion symbol with charge.

____

e. Show the stable electron configuration with long hand for fluorine ion. Which noble or inert gas is the stable fluorine ion like?

____

3. a. Show the complete or long hand electron configuration for calcium, Ca.

____

b. How many valence electrons does calcium have?_______________________

c. Will calcium tend to gain or lose electrons? ___________________________

d. What is the charge on the stable calcium ion? Show the calcium ion symbol with charge.____
e. Show the stable electron configuration with long hand for calcium ion. Which noble or inert gas is a stable calcium ion like?_____________________________

4.   a. Write the electron structures of lithium and oxygen atoms.

b. After drawing the structures of these two atoms draw their ionic structure.

5. Why helium gas is used in Goodyear blimps floating in the sky during sporting event?

6. Visit this website and explain in your words the relationship between this story and chapter 9, p 286.

http://www.npr.org/sections/krulwich/2011/10/03/140815154/dissolve-my-nobel-prize-fast-a-true-story

Experiment 9. Molecular Modeling: Lewis structures and VSEPR Theory

Pre-Lab Assignment

Name:_______________________________   **Date:**_________________

1. Draw the Lewis structures of the following compounds:

a. carbon monoxide
b. boron trichloride
c. Formaldehyde CH_2O

2. Circle the correct answer. T if true, F if it is false.

a.   All valence shell electron pairs around the central atom in a molecule repel each other. T   F
b.   All valence electron pairs around the terminal atoms in a molecule attract each other. T   F

3. Draw the Lewis structure for each of the following:

HF,

F_2,

Sulfuric acid,

Nitrate anion.

Experiment 9 Molecular Modeling: Lewis structures and VSEPR Theory

Objective:

Construct models of simple molecules of methane, ethane, propane and butane and draw Lewis structures and predict the shapes of methane, ammonia, water, and carbon dioxide.

Introduction:

In this exercise you will learn about the three-dimensional shapes of four classes of hydrocarbons: alkanes, alkenes, alkynes and an aromatic hydrocarbon, benzene. You will draw Lewis structures and construct ball-and-stick models of several compounds and', for some, you will be asked to draw their three-dimensional spatial formulas.

The chemical and physical properties of a covalent compound are a function of the composition of that compound (the atoms that compose it), the way the atoms are joined together (single, double or triple bonds) and the three-dimensional shape of the molecule. In the last 30 years of chemistry a great deal of work has gone into the discovery of the importance of molecular shapes. There are four classes of hydrocarbons: alkanes are hydrocarbons that contain only carbon-carbon single bonds; alkenes are hydrocarbons that contain at least one carbon-carbon double bond; alkynes are hydrocarbons that contain at least one carbon-carbon triple bond; and aromatic hydrocarbons are those containing at least one aromatic group, most often the benzene ring. Though there are thousands upon thousands of hydrocarbons, their three-dimensional shapes are built from only four fundamental carbon substructures that are connected together in different sequences. One of the first things you learn about the element carbon is that it always has four bonds in its compounds. This set of four bonds can be made up of four single bonds, two single bonds and one double bond.

The saturated hydrocarbons are called alkanes. They are made up of carbon-carbon single bonds. To learn about them, use of balls as representing atoms and sticks for bonds is extremely helpful. It gives the three dimensional picture of an alkane. Usually, black ball represents a carbon atom and it has four holes, red ball indicates oxygen atom and it has two holes, yellow ball is for hydrogen with one hole, blue represents nitrogen with three holes and green, orange, purple each one with one hole indicates chlorine, bromine and iodine atoms respectively.

Lewis Structures and VSPER Theory:

1. Pairs of electrons in the valence shell of a central atom repel each other.

2. These pairs of electrons tend to occupy positions in space that minimize repulsions and maximize the distance of separation between them.

3. The valence shell is taken as a sphere with electron pairs localizing on the spherical surface at maximum distance from one another.
4. A multiple bond is treated as if it is a single electron pair and the two or three electron pairs of a multiple bond are treated as a single super pair. Where two or more resonance structures can depict a molecule the VSEPR model is applicable to any such structure.

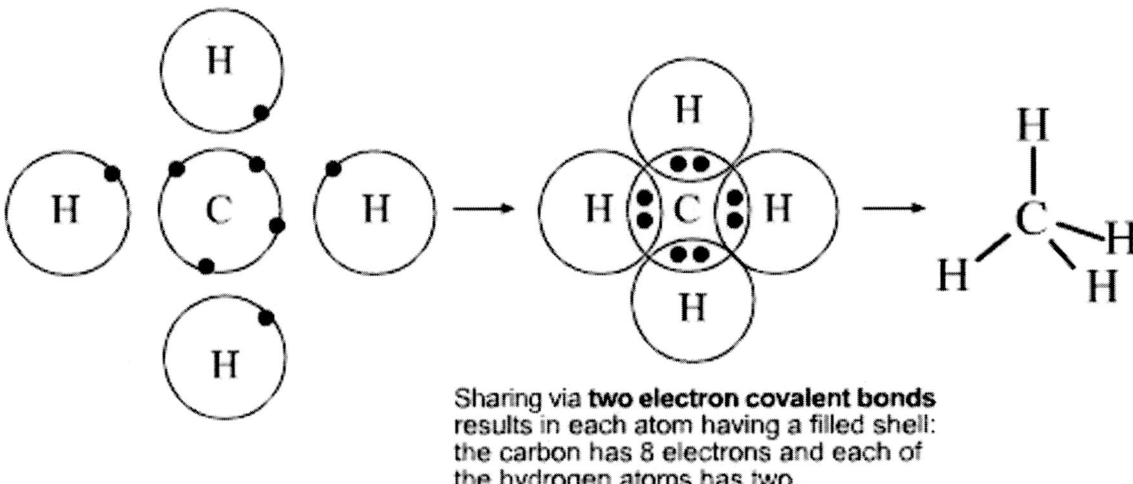

Sharing via **two electron covalent bonds** results in each atom having a filled shell: the carbon has 8 electrons and each of the hydrogen atoms has two.

<u>CH$_4$: Methane</u>

Let us write the Lewis structure of methane.

Molecular formula tells us that 'C' is minority and 'H' atoms are majority.

Total number of valence electrons:

C = 4 and each 'H' has 1, so for 4 "H" atoms, there are 4 valence electrons. Adding them gives 8 total valence electrons.

Carbon is minority and H atoms are in majority, so we can write them in the arrangement as shown below. If we connect all 'H' atoms to carbon by a line, we have used all eight electrons. No electron is left over. But 'C' is satisfied by an octet and all 'H' atoms have duet.

Second Example: Now let us write Lewis structure of ammonia: NH_3

N is minority and three 'H' atoms are majority.
N has 5 valence electrons and 3 'H' contributes 3 valence electrons. Total number of valence electrons in ammonia is 8.
N is minority and H atoms are in majority, so that we can write them in the arrangement as shown below. If we connect all 'H' atoms to nitrogen by a line, we have used six electrons. Two electrons are left over and to satisfy octet for 'N', we should keep on 'N' as unshared pair.

Third Example: Let us write Lewis structure for water (H_2O).

Minority atom is 'O' and majority atoms are 'H'. Total number of valence electrons will be eight, 6 from 'O' and 2 from two hydrogen atoms. Writing 'O' in the middle and two 'H' atoms left and right, we can connect both 'H' atoms to 'O' by a single covalent bond. We have used 4 electrons and still 4 electrons or two pairs of electrons remain. We should place both of them on oxygen that will satisfy the octet of oxygen. Both 'H' atoms have duet.

Fourth Example: Now consider carbon dioxide CO2

Carbon (C) has four valence electrons x 1 carbon =
Oxygen (O) has six valence electrons x 2 oxygen atoms =
There are a total of 8 electrons to be placed in the Lewis structure. First, connect the central atom to the other atoms in the molecule with single bonds. Carbon is the central atom, the two oxygen atoms are bound to it and electrons are added to fulfill the octets of the outer atoms. Complete the valence shell of the outer atoms in the molecule. Place any remaining electrons on the central atom. There are no more electrons available in this example. If the valence shell of the central atom is complete, you have drawn an acceptable Lewis structure. Carbon is electron deficient - it only has four electrons around it. This is not an acceptable Lewis structure. If the valence shell of the central atom is not complete, use a lone pair on one of the outer atoms to form a double bond between that outer atom and the central atom. Continue this process of making multiple bonds between the outer atoms and the central atom until the valence shell of the central atom is complete. Please check to make sure that you has used the correct number of electrons in the Lewis structure and that no atom that cannot exceed its valence shell, does not.

Procedure:

Part-A: Simple Alkanes: Methane, Ethane, Propane:

1. Get one molecular model kit for you. Take out a black ball and four small sticks. Put each stick in each hole of the black ball. Now select four yellow balls. Connect them to the black ball containing four sticks. You constructed a model of methane.

2. Draw the three dimensional shape of the methane molecule in experimental data sheet.

3. Using two black balls and a longer stick, connect them. Choose six yellow balls and six smaller sticks. Put each stick in one yellow ball. Attach three yellow balls on one black ball and the remaining three on the other black ball.

4. Now you have a model of ethane (C_2H_6).

5. Draw the three dimensional picture of it.

Using same procedure, described in the first ten steps, make a model of propane with three black balls and eight yellow balls. You have to use three longer sticks and eight smaller sticks. Draw the three dimensional picture of it.

Using same procedure, described, make a model of butane with four black balls and ten yellow balls. You have to use four longer sticks and ten smaller sticks. Draw the three dimensional picture of it.

methane CH_4	H | H—C—H | H	
ethane C_2H_6	H H | | H—C—C—H | | H H	
propane C_3H_8	H H H | | | H—C—C—C—H | | | H H H	
butane C_4H_{10}	H H H H | | | | H—C—C—C—C—H | | | | H H H H	

1. Look at the molecular formula and determine which atoms are few and which are in the majority.

2. Count the total number of valence electrons in all the atoms of the molecule. If you are writing Lewis structure of a polyatomic ion, adjust the number of electrons for the charge.

3. Placing the minority atom in the middle, surround it by majority atoms. If you can not distinguish between minority and majority, use the connecting pattern given in the molecule.

4. Connect atoms to the minority atom in the center by a line. A line represents a single covalent bond and involves use of two electrons.

5. Subtract number of electrons used in lines from the total, counted in step #2.

6. Use remaining electrons, so that each atom has eight electrons, except H, which needs a duet.

7. If electrons are less than required to provide each atom octet, the atoms must share more pairs to form multiple bonds.

Experiment 9 Molecular Modeling: Lewis structures and VSEPR Theory

Lab Report:

Name_________________________   **Date:**____________

A: <u>**Three Dimensional Drawing:**</u>   **Lewis Structure**

   1. Methane CH_4

   2. Ethane CH_3-CH_3

   3. Propane CH_3-CH_2-CH_3

   4. Butane CH_3-CH_2-CH_2-CH_3

B. **VSEPR Theory Report Sheet**

Procedure: Take appropriate balls and sticks and make the models of ammonia, water and carbon dioxide. To make multiple bonds, use metal springs. Then answer by filling the blanks in the table below as shown for methane in the first column.

	Methane, CH_4	Ammonia, NH_3	Carbon dioxide	Water
Number of valence electrons in the molecule		_______	_______	_______
Number of valence electrons around central atom in the molecule		_______	_______	_______
# of single bonds on central atom		_______	_______	_______
# of lone pairs on central atom		_______	_______	_______
# of double bonds on central atom		_______	_______	_______
Predict shape		_______	_______	_______

<u>**Lab # 10-12 Gas Laws: a. Boyle's Law, b. Charles' Law and c. Gay Lussac's Law**</u>

<u>**Pre Lab Assignment**</u>

Name:_______________   **Date:**__________

<u>**Questions:**</u>

1. A cylinder contains 5.0 liter of oxygen gas at a pressure of 850 mm. If the pressure is increased to 1234 mm, what will be the new volume of oxygen gas?

Answer:_____________________

2) A sample of He gas has a volume of 5.8 L at 700 mm at room temperature. What will be its volume if the pressure is changed to standard pressure? Assume the temperature remains constant.

Answer:_____________________

3) What is the new volume of N_2 gas in a cylinder that initially holds at constant temp at 380 torr pressure and has a volume of 520 mL, then pressure is changed to 760 torr.

   Answer:_____________________

4) An auto has a cylinder with a volume of gasoline and air mixture of 2.0 L at 1000mm pressure. If the volume expands to 2.35 L, what will be the final pressure of gasoline and air mixture in mm inside the auto engine?

Answer:_____________________

5). A sample of neon gas has a volume of 50.0 L at $25^0\,C$ and 723 torr. What volume will the sample have if the pressure is changed to 1.4 atm.? Assume the temperature is kept constant.

Answer:_____________________

6). What will be the final volume of a gas at 0.5 atm. and 1.23 L. if the pressure was changed to 1 atm. at the same temperature?

Answer:_____________________

7) What is Gay-Lussac's Law?

8) What is the relationship in each case?
a) Pressure and volume_______________________________
b) Pressure and temperature___________________________
c) Pressure and moles_______________________
d) Volume and temperature_____________________________

9). Why do we need to learn about gas laws and how do the gases behave?

10) State Charles' Law in words and mathematical expression.

Lab # 10-12 Gas Laws: a. Boyle's Law, b. Charles' Law and c. Gay Lussac's Law

Objective:

a. To understand the relationship between the pressure and volume of a gas at constant temperature as stated in Boyle's Law, using gas pressure sensor with computer interface.
b. To measure the volume of a fixed quantity of air as the temperature changes at constant pressure.
c. Study the relationship between the temperature of a gas sample and the pressure it exerts and determine from the data and graph, the mathematical relationship between the pressure and absolute temperature of a confined gas.

Introduction:

In 1660, Irish scientist Robert Boyle studied the compressibility of gases at constant temperature. He noted that the volume of a given sample of gases is inversely proportional to the pressure at constant temperature. For example, if we double the pressure on a volume of fixed mass of gas, at constant temperature, the volume will be halved.

V is proportional 1/P.

P is proportional to 1/V.

The above relationship may be expresses as

$$PV = k$$

$$P\ i_{initial}\ X\ V\ i_{initial}\ =\ P\ _{final}\ X\ V\ _{final}$$

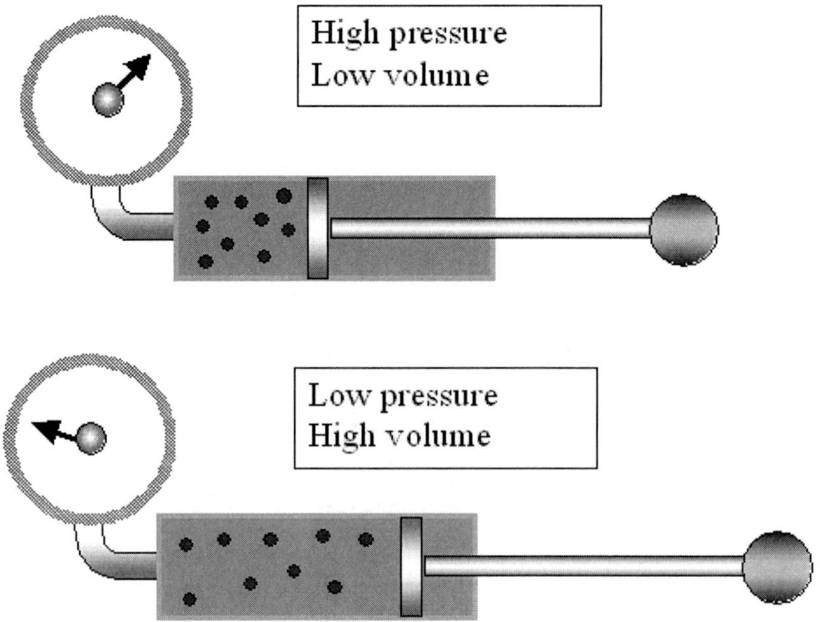

The product of pressure and volume remains constant, if the temperature remains unchanged. If 50 mL of oxygen gas is compressed from 20 atm of pressure to 40 atm of pressure, *what is the new volume at constant temperature?* Simply set up a data table to identify the relevant variables.

First, let's rewrite the above question, identifying the variables: If 50 mL (V_1) of oxygen gas is compressed from 20 atm (P_1) of pressure to 40 atm (P2) of pressure, what is the new volume (V_2) at constant temperature? Plug the above values into the equation $P_1 V_1 = P_2 V_2$ 20 atm X 50 mL = 40 atm X V_2 Solving for V_2, we get 25 mL as our answer.

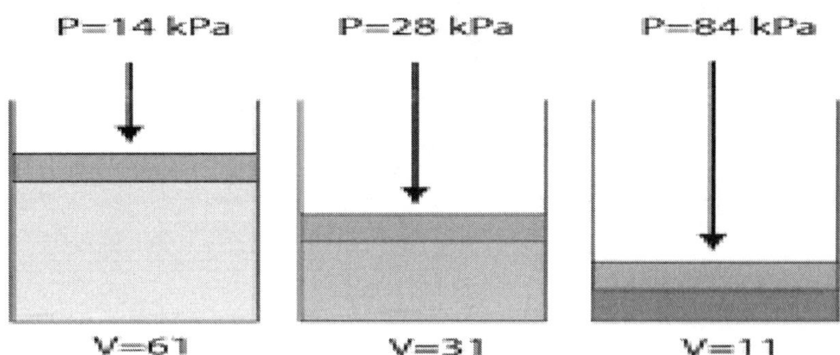

<u>Example</u>: *A sample of helium gas has a volume of 5.8 liter at 700 mm at room temperature. What will be its volume if the pressure is changed to standard pressure and temperature remains constant?*

<u>Solution:</u>

Let us write down all known values and "x" for unknown value.

P (initial) = 700 mm   V (initial) = 5.8 L

P (final) = 760 mm   V (final) = x

$P_{initial}$ X $V_{initial}$ = P_{final} X V_{final}

700 mm   X 5.8 L   =   760 mm X   x

$$\frac{700 \text{ mm} \quad \text{x} \quad 5.8 \text{ L}}{760 \text{ mm}} = \text{x}$$

5.34 L =   x

Boyle studied what happened to the volume of the gas in the sealed end of the tube as he added mercury to the open end.

For example, if we double the pressure on a volume of fixed mass of gas, at constant temperature, the volume will be halved.

$$P_{initial} \text{ X} \quad V_{initial} = P_{final} \text{ X} \quad V_{final}$$

The product of pressure and volume remains constant, if the temperature remains unchanged. Boyle noticed that the product of the pressure times the volume for any measurement in this table was equal to the product of the pressure times the volume for any other measurement, within experimental error.

$$P_1V_1 = P_2V_2$$

The pressure of a gas depends on the volume. At constant temperature, the pressure of a gas varies inversely to its volume. This is called Boyle's law. By rearranging the relationship, it can be shown that the product of any volume to its corresponding pressure remains constant. According to Boyle's law, volume decreases when the pressure increases and as the pressure decreases, the volume of the gas increases.

Charles' Law:

In 1787, Jacques Charles, a French scientist studied the relationship between volume and temperature of a gas at constant pressure. The temperature scale, he used was absolute temperature scale, called kelvin and he found that the volume of a fixed mass of a gas is directly proportional to absolute temperature, if the pressure is kept constant.

V is directly proportional to T, where V stands for volume and T represents temperature in Kelvin scale.

$$V = k\,T$$

$$V/T = k$$

$$\frac{V_{initial}}{T_{initial}} = \frac{V_{final}}{T_{final}}$$

Example:

An automobile engine has a cylinder with a volume of gasoline and air mixture, 2.0 L at 180 c. if the volume expands to 2.35 L, what will be the temperature of gasoline and air mixture inside the engine?

Solution:

Here temperature of the content in the engine is changing as the volume expands. The application of Charles' law will help us to find new temperature, which will be higher than 180 c.

V (initial) = 2.0 L, V (final)= 2.35 L, T(initial) = 180 +273= 453 kelvin

$$\frac{V\ (initial)}{T\ (initial)} = \frac{V\ (final)}{T\ (final)}$$

$$\frac{2.0\ L}{453\ K} = \frac{2.35\ L}{x}$$

Therefore

$$\frac{2.35\ L\ x\quad 453\ K}{1.0\ L} = 532.3\ K$$

To convert to degree centigrade:

$$(523.3K - 273) = 259.3^0\ c$$

The solution shows higher temperature than 180 c and it makes sense according to Charles' Law.

Gay-Lussac's Law :

Gases are made up of molecules that are in constant motion. Due to their collision with the walls of their container, gases exert pressure. The velocity and the number of collisions of these molecules are affected when the temperature of the gas increases or decreases. In this experiment, you will study the relationship between the temperature of a gas sample and the pressure it exerts. Pressure and temperature data pairs will be collected during the experiment and then analyzed. From the data and graph, you will determine what kind of mathematical relationship exists between the pressure and absolute temperature of a confined gas.

Gay-Lussac's Law, or the Pressure Law, was found in 1809. It states that, for a given mass and constant volume of an ideal gas, the pressure exerted on the sides of its container is directly proportional to its absolute temperature.

As a mathematical equation, Gay-Lussac's Law is written as either:

$P \propto T$, or

P1/T1 = P2/T2

Where P is the pressure, T is the absolute temperature, and k_3 is proportionality constant

Procedure:

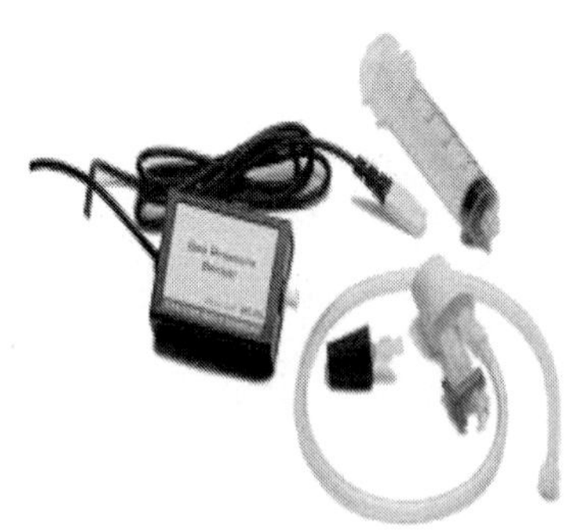 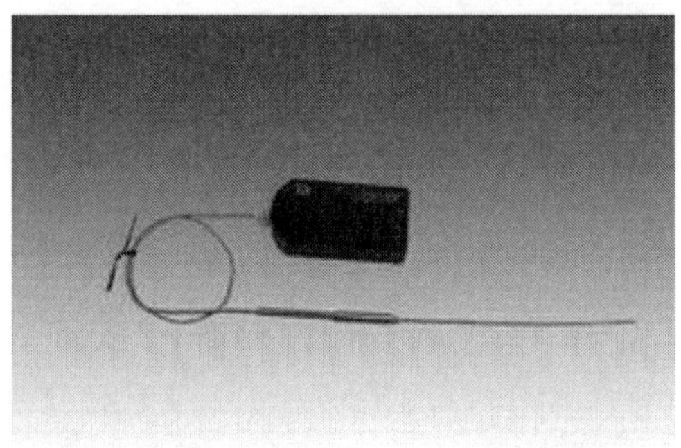

a. Boyle's Law:
1. Collect the following items for the experiment: a computer, 20 mL gas syringe, gas pressure sensor and a computer interface, printer.
2. Observe carefully and locate channel 1 on the computer interface. Plug the gas pressure sensor into channel 1 of the interface.
3. Get a 20mL gas pressure syringe, move the piston of the syringe to 10mL position. Attach it to the valve of the Gas Pressure Sensor.
4. Turn on the computer and open chemistry experiment file 06 Boyle's Law from Logger Pro.
5. To account for the extra volume of the trapped air in the syringe, you will need to add 0.8 mL to volume reading from the syringe. So 7.0 mL of syringe volume must be recorded as 7.8 mL
6. Click "COLLECT" to begin the process.
7. Record pressure as well as volume.
8. Move the piston to a new position, instead of 7.0 mL, for example 8.0 mL and hold tight. Record pressure.
9. Continue the procedure for syringe volume of 9.0, 10.0, 11.0 13.0, 15.0, 17.0, 18.0 and 19.0 mL
10. Click STOP.
11. Choose INVERSE mathematical relationship and click Curve Fit, then choose "Variable Power" $y=Ax^n$ from the list at the lower left side of the screen. Enter power value n in the Power edit box -1. Click TRY FIT. Print.

<u>**b. Charles' Law:**</u>

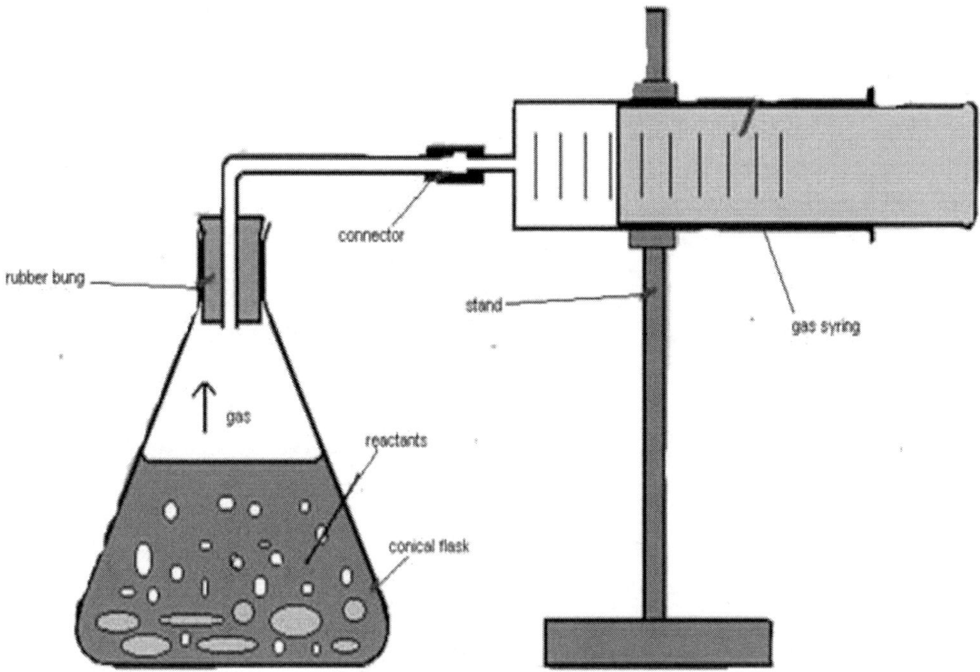

1. Heat a 600-ml of water to boiling
2. Open the end of the syringe and set the plunger at the 30-ml mark. Close the end of the syringe. Keep the syringe on the ring stand and place the beaker of hot water under it.
3. Adjust the height of the clamp holding the syringe so that the syringe is as far down in the beaker as possible. Wait about 2 minutes for the air in the syringe to come to the same temperature as the water.
4. Record the water temperature and the volume on the syringe on your data sheet.
5. Raise the clamp and syringe out of the water. Replace the beaker of hot water with a beaker of hot tap water (or add ice to the boiling water). Lower the syringe into the water and wait 2 minutes for equilibration to occur.
6. Record the temperature and the volume on the syringe.
7. Remove the clamp and syringe from the water. Replace the beaker of warm water with cold water. Wait 2 minutes and then record the temperature and the volume on the syringe. Pour off about half of the water and fill the beaker with ice. Place the syringe in the water and record the temperature and volume as before.

<u>**c. Gay Lussac's Law:**</u>

Collect the sensor, probe, tubes, rubber stoppers and computer provided. Take one liter beakers and fill up with 700 ml water. Heat them till it boils. This is boiling-water bath and it will be used for second reading and data collection. Take about 700 mL of room-temperature water into a second 1 L beaker. Add ice in it. This is cold water bath and will be used for the first reading.

<u>COLD WATER READING:</u>

1. This is first part to study pressure and temperature relationship at cooler temp.
2. Take 1 liter beaker with ice in it.
3. Collect computer, sensors, material and equipment provided.
4. Turn on the computer and prepare the Temperature Probe and Gas Pressure Sensor for data collection.
5. Connect the Gas Pressure Sensor to Channel 1 of the computer interface and the Temperature Probe to Channel 2 of the computer interface
6. You are provided a rubber-stopper assembly with a piece of heavy-wall plastic tubing connected to one of its two valves. Attach the connector at the free end of the plastic tube to the open stem of the Gas Pressure Sensor as shown by the instructor.
7. Leave its two-way valve on the rubber stopper open till instructed.
8. Put the rubber-stopper assembly into a small Erlenmeyer flask. Make sure the rubber stopper fits tight.
9. Close the 2-way valve above the rubber stopper. Turn the valve handle making it perpendicular as explained. The air sample is now trapped in the flask.
10. Prepare the computer for data collection by opening the file called" Pressure and Temp."
11. Click "COLLECT" to begin data collection.
12. Collect pressure *vs.* temperature data for your gas sample by the following three steps:

i. Place the flask into the ice-water bath. Make sure the entire flask is covered as demonstrated. Stir to dissolve ice.
ii. Place the Temperature Probe into the ice-water bath.
iii. When the pressure and temperature readings stabilize, click "KEEP. You have now saved the first pressure-temperature data and you can print it.

13. Repeat the procedure of step 12 using <u>the hot water bath.</u>
14: Use a ring stand and utility clamp; suspend the temperature probe in the boiling-water bath, as demonstrated. After the temperature probe has been in the boiling water for a minute, place the flask into the boiling-water bath and repeat procedure in the above box of step 12.

15. Click STOP when you have finished collecting data. Record the pressure and temperature values and print a copy of the table.
16. Examine your graph of pressure *vs.* temperature (°C). In order to determine if the relationship between pressure and temperature is direct or inverse, use an absolute temperature scale, the Kelvin absolute temperature scale. Click on the horizontal-axis label, select "Temp Kelvin" to be displayed on the horizontal axis. <u>"Autoscale"</u> both axes starting with zero, double-click in the center of the graph to view Graph Options, click the Axes Options tab, and select <u>"Autoscale"</u> from 0 for both axes.
17. Click Curve Fit, then choose "Variable Power" <u>$y=Ax^n$</u> from the list at the lower left side of the screen. Enter power value n in the Power edit box -1. Click TRY FIT. Print.

<u>Lab # 10-12 Gas Laws: a. Boyle's Law, b. Charles' Law and c. Gay Lussac's Law</u>

Name:_________________________   Date:_________________________________

<u>Lab Report:</u>

<u>Experimental Data:</u>_________

<u>a. Boyle's Law</u>

1. Attach your graph and check each value of volume, which you have selected in the procedure earlier. Write the corresponding pressure. Multiply pressure and volume to find K. For example:
If volume was 7.8 and pressure was 142.65, your K will be 7.8 x 142.65=1112.67.

<u>Pressure</u>   **<u>Volume</u>**   **<u>Pressure x Volume</u>**

<u>1.</u>
<u>2.</u>
<u>3.</u>
<u>4.</u>
<u>5.</u>
<u>6.</u>
<u>7.</u>

b. Charles Law:

Temp. o C	Temp. o K	Volume, ml
1.		
2.		
3.		
4.		

1. Using the computer provided, on a spread sheet plot the above data with T on the X-axis and V on the Y axis. Draw a straight line which best fits the data points.
2. Are the data points reasonably close to the straight line? If not, can you think of a reason that Charles' law was not confirmed?

<u>**c. Gay Lussac's Law:**</u>
Attach table and graph here. Write a conclusion of the experiment.

<u>**Lab. 13.**</u>   <u>**Liquids and Intermolecular Forces**</u>

Prelab:

Name:_______________________________ Date:___________________________
<u>**PRE-LAB EXERCISE**</u>

Prior to doing the experiment, complete the Pre-Lab.

1. Draw a structural formula for a molecule of each compound to be used in this experiment.

2. Now determine the molecular weight of each of the molecules.

3. Study the following table:

<u>**Liquid**</u>	<u>**Formula**</u>	<u>**Molar mass**</u>	<u>**Boiling Point**</u>
Water	**H-O-H**	**18**	**100^0 C**
Alcohol	**C_2H_6O**	**46**	**78^0 C**
Ether	**C_2H_6O**	**46**	**36^0 C**

As a general rule, as the molar mass increases, the boiling point should increase too. But in the above cases, water with the lowest molar mass has the highest boiling point. Why do we see such a contradictory result? Explain.

Lab 13. Liquids and Intermolecular Forces

OBJECTIVES:

1. Study temperature changes caused by the evaporation of several liquids.
2. Relate the temperature changes to the strength of intermolecular forces of attraction.

Introduction:

Intermolecular forces and molecular weight determine the vapor pressure of a liquid. Strong attractive intermolecular forces decrease the escaping ability of liquid molecules. The stronger and powerful attractive forces keep the liquid molecules holding each other. That lowers the vapor pressure.

Smaller molecules are capable of escaping faster and with ease from the surface of the liquid, compare to larger and bigger molecules with higher formula weight. Lower the molecular weight; higher will be the vapor pressure. Similarly, higher the molecular weight, lower will be the vapor pressure. What is the effect of temperature on the vapor pressure of a liquid?

Vapor pressure increases as the temperature of the liquid goes up. This occurs due to the facts that raise in temperature increases the kinetic energy of the molecules. Higher the kinetic energy, higher will be the escaping ability of the molecules from the surface of the liquid.

Evaporation and intermolecular attractions will be studied in this experiment. Temperature Probes are placed in various liquids. Evaporation occurs. When the probe is used, the evaporation which is an endothermic process can be correlated with temperature decrease. The magnitude of a temperature decrease is related to the strength of intermolecular forces of attraction.

The results can be used to predict, the temperature change for several other liquids.

The three alkanes are n-pentane, C_5H_{12}, n-hexane, C_6H_{14} and n-heptane C_7H_{16}. The three alcohols also contain the -OH functional group. Ethanol, C_2H_5OH, Propanol and isopropanol are used in this experiment. You will study the molecular structure of alkanes and alcohols for the presence and relative strength of two intermolecular forces like hydrogen bonding and dispersion forces.

PROCEDURE

1. Make sure you have collected the following materials and equipment, before you begin the experiment: laptop computer and accessory, ethanol (ethyl alcohol), 1-propanol, isopropanol, two temperature probes, filter papers, n-pentane, n-hexane, n-heptane, small rubber bands, masking tape NOTE: The compounds used in this experiment are flammable organic. Avoid inhaling them.

2. Connect the probes to the computer interface. Prepare the computer for data collection by opening the file called "Evaporation" from the Chemistry folder.

3. Wrap Probe 1 and Probe 2 with filter paper secured by small rubber bands as shown by the instructor. Roll the filter paper around the probe tip. Please slide the rubber band up on the probe, and wrap the paper around the probe. Now slip the rubber band over the wrapped paper.

4. Stand Probe 1 in the ethanol container and Probe 2 in the 1-propanol container.

5. Prepare 2 pieces of masking tape to tape the probes later.

6. After the probes have been in the liquids for at least 30 seconds, begin data collection by clicking COLLECT. Monitor the temperature for 15 seconds to establish the initial temperature of the liquid. Then simultaneously remove the probes from the liquids.

7. When both temperatures have reached minimums and have begun to increase, click STOP data collection. Click the Statistics then click to display a box for both probes. Record the maximum (t1) and minimum (t2) values for Temperature 1 (ethanol) and Temperature 2 (1-propanol).

8. For each liquid, subtract the minimum temperature from the maximum temperature. This is the temperature change during evaporation.

9. Now, predict the size of the temperature difference value for isopropanol. Record your prediction.

10. Test your prediction by repeating earlier procedure using isopropanol for Probe 1 and n-pentane for Probe 2.

11. Based on the temperature difference values for all four substances, predict the temperature difference values for n-hexane and n-heptane. Record your predicted value.

12. Test your prediction, using isopropanol with Probe 1 and n-heptane with Probe 2.

Plot a graph of temperature values of the three alcohols as well as three alkanes versus their respective molecular weights.

EXPERIMENTAL DATA:

Observed from experiment:

Substance (°C)	t1 (°C)	t2 (°C)	(t1–t2)
ethanol			
n-propanol			
isopropanol			
n-pentane			
n-hexane			
n-heptane			

Predicted value for:

Isopropanol			
n-heptane			

Questions:

1. Attach your printed graph.

2. **Circle the correct answer.**

1) The hydrogen bonding will not exit between molecules of

a) H_2O
b) NH_3
c) H_2
d) HF

2) The high boiling point and other unusual properties of water are due to

a) hydrogen bonding
b) oxygen bonding
c) covalent bonding
d) none of the above

3) The hydrogen bonding will not exit between molecules of

a) CH_3CH_2OH
b) NH_3
c) BeH_2
d) CH_3OH

4) The iceberg floats on water are due to which property of water

1. Hydrogen bonding
2. Oxygen bonding
3. Covalent bonding
4. None of the above

14. Review for Lab Final

Sample Questions:

Ref: Tro's Text page: 6 Read for #1.
1. What is scientific method? What is the application of it?

Ref: Tro's Text page#: 24 and 36 Read for #2.

2.
a) Who is taller, you (5feet 8 inches) or your laboratory partner (1.59 meters) and by how many decimeters?

b) Did you use the same balance during the mass determination of evaporating dish and medicine dropper? Why?

c). A metal block having mass of 14.55 grams was added to a graduate cylinder. The water level in the graduate cylinder was 20.00 ml, but due to metal block inside, it went up to a new level of 24.00 ml. Calculate the density of the metal block.

d).What is the mass of an unknown object, which has a density of 0.775 g/ml and a volume of 51.0 ml? What do you think about the state of the unknown object? Was it solid or liquid?

e) A student was asked to find the density of a metal block. First she weighed it and recorded the mass as 26.0 g. She took a graduated cylinder and placed water up to 60-ml marker. When she put the metal block in the graduated cylinder, the level went up to 65 ml. What is the density of this metal block?

Ref: Tro's Text page#: 63 Read for #3.

3.
a) Imagine your friend's house has old paints. He suspects that it contains lead and asks your advice to check out. If all the chemical reagents are available, which one will you select? Why?

b) The tap water in a home, near a contaminated site, turns deep red color with KSCN. Which heavy metal cation may be responsible? Which cation is the least reactive, among all the tested cations? Consider two cations give similar results with all the detecting reagents, how do you solve this problem to identify the unknown cation?

c) Why is it important to use separate capillary tubes, one for each cation? Why is it important to keep solution spot with the capillary as small as possible? Why do we cover the chromatography chamber or beaker with a watch glass or aluminum foil? What kind of change is responsible when DMG comes in contact with nickel, physical or chemical? Why can we use one reagent, say HCl for identification of all three cations? What is the best reagent to detect copper cation? Which is the ideal for nickel? Which one will you use for iron? Explain your reasoning in each case.

Ref: Tro's Text page#: 63 Read for #4.
4.
a. A chemistry student started with 2.0 g of copper powder. He added nitric acid to it and a toxic brown gas evolved. He left the laboratory to avoid the fumes and when he came back it was a bluish green solution. He followed the procedure and at the end of two weeks, he recovered 1.498 grams of copper. Answer the following questions:

How much % copper did he recover?
What must be the reasons for less than 100 % recovery of the copper?

b. If an unknown sample contains 1.00 g mixture. If a student separates the three pure substances and records the results as below:
a) Amount of ammonium chloride = 0.28 g
b) Amount of silicon dioxide = 0.41 g
c) Amount of sodium chloride = 0.20 g

What is the percentage recovery of matter?

c. What is the difference between sublimation and evaporation?

d. If a student had 90% recovery of all the three components combined together and the mass of the sample given was 2.995 g, how many grams did she recover

e. Mr. Fed Brose was given a mixture of three solids. He separated them and got the following data. When he calculated % recovery, it came out 91 %. Using the following information, find out the mass of original sample given to Mr. Max Smith.

Ammonium chloride: 0.8 g.
Sodium chloride : 0.93 g
Silicon dioxide: 1.0 g.

Explain the potential source of errors.

<u>Ref: Tro's Text page#: 77 Read for #5.</u>

<u>5.</u>

<u>a.</u> If 50 grams of iron is heated to 100^0 C and placed in 500 grams of cold water, which is at temperature 0^0 C, what will be the final temperature of water and iron?
Use: Specific heat of iron 0.106 cal/g c and specific heat of water 1 cal/g c.

b. When a piece of an alloy with mass 13.0 grams at temperature of 460^0 C was dropped in water, its temperature cooled down to 48.92^0 C. The amount of water was 740 grams. It was at room temperature (29^0 C). If the specific heat of water is 4.18 Joules/gram ^{0}C, calculate the specific heat capacity of unknown alloy.

c. How much heat is released when 90.00 grams of a metal is cooled from 99.6°Cto 29.1°C? The specific heat of the metal is 0.481 J/g·^{0}C.

<u>d</u>:A hot piece of zinc metal, at 3400 ^{0}C having a mass of 15.238 g, is placed in ice cold water at zero degree centigrade. What is the final temperature of the zinc in water, if the specific heat of water is 4.18 Joules/gram ^{0}C and the specific heat of zinc is 0.388 J/g·^{0}C.

<u>e</u> An iron skillet weighing 15 kg is heated on a stove to 178.3°C. Suppose the skillet is then cooled to room temperature, 25°C. How much heat energy is lost by the iron in the cooling process? The specific heat of iron is 0.450 J/g·^{0}C.

<u>f.</u> How much heat is released when 90.5 g of a metal is cooled from 91.6°Cto 20.5°C? The specific heat of the metal is 0.481 J/g·^{0}C.

<u>Ref: Tro's Text page#: 182-183 Read for #6.</u>

6.
a) Explain the difference between the hydrate and the anhydrous salt in terms of mass and structure.

b) A hydrate with formula SrI2.x H2O weighs 16.20 grams before heating. On heating, water was completely lost and the weight of anhydrous salt was found to be 12.30 grams. Calculate the value of x in SrI2.x H2O.

c) Calculate the number of grams in 1.5 moles of MgSO4 and 2.35 moles of water.

d) What would have been the result if you had heated 20.0 grams of the hydrate, instead of 2.0 grams?

e) A hydrate weighing 6.02 g of CuSO4.X H2O is heated in a crucible. After cooling to room temperature, the mass was found to be only 5.69 g. Calculate the number of moles of water lost in the experiment as well as % of water..

f). Use the following data and calculate value of X in a hydrate which is $MgSO_4:\mathbf{X}\ H_2O$, where X is number of moles of water in the hydrate.

Mass of crucible   = 30.6 g
Mass of crucible + Hydrate   =33.22 g
Mass of crucible + product after heating = 31.22g.

<u>Ref: Tro's Text page#: 204-239, Read for #7.</u>

7.
a. Balance the following equation: ___ $NaHCO_3$ = ___ Na_2CO_3 + ___ CO_2 + ___ H_2O.
What is the sum of the coefficients for

i) $NaHCO_3$ and Na_2CO_3 only?
ii) CO_2 and H_2O <u>only</u>?
iii) What is the sum of all the coefficients?

b. Balance the following equation: ___ $AlCl_3$ + ___ $Pb(NO_3)_2$ = ___ $PbCl_2$ + __$Al(NO_3)_3$
What is the sum of the coefficients for $AlCl_3$ and $Pb(NO_3)_2$ <u>only</u>?

c. Balance the following equation: ___ Mg_3N_2 + ___ H_2O = ___ MgO + ___ NH_3
What is the sum of all the coefficients?

d. What is the balanced chemical equation for the following process?

A solution of sodium hydroxide reacts with a solution of chromium(III) chloride, producing a solution of sodium chloride and chromium(III) hydroxide.

<u>**Ref: Tro's Text page#: 302, Read for #8.**</u>

8. (a) Which atom(s) has unpaired electron(s) among, carbon, nitrogen, oxygen and fluorine? How many unpaired electron(s)? List all possible answers. You are cooking soup with table salt and boils over on the stove. It burns with a bright golden yellow flame, what must be the reason?

(b) Which is the smallest alkali metal? Which is the smallest halogen element? Why were you able to identify the unknown solution?

C) You are cooking soup with table salt and boils over on the stove. It burns with a bright golden yellow flame, what must be the reason?

Ref: Tro's Text page#: 324-349, Read for #9.

9.

A. What is the shape of CO_2 molecule?

B. Write the Lewis structure of sulfite and sulfate ions.

C. Predict the shape of sulfite and sulfate ion.

Ref: Tro's Text page#: 365, 371 and 380, Read for #10.

10.

a) Questions: In Boyle's Gas Law a) Multiply the value of total pressure and corresponding volume for reading 1 and 2. Did you get the same value? Why?

b) A cylinder contains 5.0 liter of oxygen gas at a pressure of 850 mm. If the pressure is increased to 1234 mm, what will be the new volume of oxygen gas?

c) An automobile engine has a cylinder with a volume of gasoline and air mixture, 2.0 L at 180 c. if the volume expands to 2.35 L, what will be the temperature of gasoline and air mixture inside the engine?

d) What is Gay Lussac's Law? State and explain in your words.

Temperature [° C]	Density [g/mL]	Temperature [° C]	Density [g/mL]
13.0	0.99987	28.0	0.99626
14.0	0.99999	29.0	0.99598
15.0	0.99973	30.0	0.99568
16.0	0.99963	35.0	0.99406
17.0	0.99953	40.0	0.99225
18.0	0.99941	45.0	0.99024
19.0	0.99927	50.0	0.98807
20.0	0.99913	55.0	0.98572
21.0	0.99897	60.0	0.98323
22.0	0.99880	65.0	0.98058
23.0	0.99862	70.0	0.97779
24.0	0.99843	75.0	0.97487
25.0	0.99823	80.0	0.97182
26.0	0.99802	85.0	0.96864
27.0	0.99780	90.0	0.96534
28.0	0.99757	95.0	0.96192

PERIODIC TABLE SHOWING HEAVY ELEMENTS AS MEMBERS OF AN ACTINIDE SERIES

Arrangement by Glenn T. Seaborg, 1945

1	2	3	4	5	6	7	8	9	10	11	12	13	14	15	16	17	18
1 H 1.008																1 H 1.008	2 He 4.003
3 Li 6.940	4 Be 9.02											5 B 10.82	6 C 12.010	7 N 14.008	8 O 16.000	9 F 19.00	10 Ne 20.183
11 Na 22.997	12 Mg 24.32	13 Al 26.97										13 Al 26.97	14 Si 28.06	15 P 30.98	16 S 32.06	17 Cl 35.457	18 A 39.944
19 K 39.096	20 Ca 40.08	21 Sc 45.10	22 Ti 47.90	23 V 50.95	24 Cr 52.01	25 Mn 54.93	26 Fa 55.85	27 Co 58.94	28 Ni 58.69	29 Cu 63.57	30 Zn 65.38	31 Ga 69.72	32 Ge 72.60	33 As 74.91	34 Se 78.96	35 Br 79.916	36 Kr 83.7
37 Rb 85.48	38 Sr 87.63	39 Y 88.92	40 Zr 91.22	41 Cb 92.91	42 Mo 95.95	43	44 Ru 101.7	45 Rh 102.91	46 Pd 106.7	47 Ag 107.880	48 Cd 112.41	49 In 114.76	50 Sn 118.70	51 Sb 121.76	52 Te 127.61	53 I 126.92	54 Xe 131.3
55 Cs 132.91	56 Ba 137.36	57 LA / 58-71 SEE La SERIES	72 Hf 178.6	73 Ta 180.88	74 W 183.92	75 Re 186.31	76 Os 190.2	77 Ir 193.1	78 Pt 195.23	79 Au 197.2	80 Hg 200.61	81 Tl 204.39	82 Pb 207.21	83 Bi 209.00	84 Po	85	86 Rn 222
87	88 Ra	89 AC / SEE Ac SERIES	90 Th	91 Pa	92 U	93 Np	94 Pu	95	96								

LANTHANIDE SERIES

57 La 138.92	58 Ce 140.13	59 Pr 140.92	60 Nd 144.27	61	62 Sm 150.43	63 Eu 152.0	64 Gd 156.9	65 Tb 159.2	66 Dy 162.46	67 Ho 163.5	68 Er 167.2	69 Tm 169.4	70 Yb 173.04	71 Lu 174.99

ACTINIDE SERIES

89 Ac	90 Th 232.12	91 Pa 231	92 U 236.07	93 Np 237	94 Pu	95	96							

127

Elements of the periodic table:

No.	Atomic Weight	Name	Sym.
1	1.0079	Hydrogen	H
2	4.0026	Helium	He
3	6.941	Lithium	Li
4	9.0122	Beryllium	Be
5	10.811	Boron	B
6	12.0107	Carbon	C
7	14.0067	Nitrogen	N
8	15.9994	Oxygen	O
9	18.9984	Fluorine	F
10	20.1797	Neon	Ne
11	22.9897	Sodium	Na
12	24.305	Magnesium	Mg
13	26.9815	Aluminum	Al
14	28.0855	Silicon	Si
15	30.9738	Phosphorus	P
16	32.065	Sulfur	S
17	35.453	Chlorine	Cl
18	39.948	Argon	Ar
19	39.0983	Potassium	K
20	40.078	Calcium	Ca
21	44.9559	Scandium	Sc
22	47.867	Titanium	Ti
23	50.9415	Vanadium	V
24	51.9961	Chromium	Cr
25	54.938	Manganese	Mn
26	55.845	Iron	Fe
27	58.9332	Cobalt	Co
28	58.6934	Nickel	Ni
29	63.546	Copper	Cu

30		65.39	Zinc	Zn
31		69.723	Gallium	Ga
32		72.64	Germanium	Ge
33		74.9216	Arsenic	As
34		78.96	Selenium	Se
35		79.904	Bromine	Br
36		83.8	Krypton	Kr
37		85.4678	Rubidium	Rb
38		87.62	Strontium	Sr
39		88.9059	Yttrium	Y
40		91.224	Zirconium	Zr
41		92.9064	Niobium	Nb
42		95.94	Molybdenum	Mo
43	*	98	Technetium	Tc
44		101.07	Ruthenium	Ru
45		102.9055	Rhodium	Rh
46		106.42	Palladium	Pd
47		107.8682	Silver	Ag
48		112.411	Cadmium	Cd
49		114.818	Indium	In
50		118.71	Tin	Sn
51		121.76	Antimony	Sb
52		127.6	Tellurium	Te
53		126.9045	Iodine	I
54		131.293	Xenon	Xe
55		132.9055	Cesium	Cs
56		137.327	Barium	Ba
57		138.9055	Lanthanum	La
58		140.116	Cerium	Ce
59		140.9077	Praseodymium	Pr
60		144.24	Neodymium	Nd
61	*	145	Promethium	Pm
62		150.36	Samarium	Sm
63		151.964	Europium	Eu
64		157.25	Gadolinium	Gd
65		158.9253	Terbium	Tb
66		162.5	Dysprosium	Dy
67		164.9303	Holmium	Ho

68		167.259	Erbium	Er
69		168.9342	Thulium	Tm
70		173.04	Ytterbium	Yb
71		174.967	Lutetium	Lu
72		178.49	Hafnium	Hf
73		180.9479	Tantalum	Ta
74		183.84	Tungsten	W
75		186.207	Rhenium	Re
76		190.23	Osmium	Os
77		192.217	Iridium	Ir
78		195.078	Platinum	Pt
79		196.9665	Gold	Au
80		200.59	Mercury	Hg
81		204.3833	Thallium	Tl
82		207.2	Lead	Pb
83		208.9804	Bismuth	Bi
84	*	209	Polonium	Po
85	*	210	Astatine	At
86	*	222	Radon	Rn
87	*	223	Francium	Fr
88	*	226	Radium	Ra
89	*	227	Actinium	Ac
90		232.0381	Thorium	Th
91		231.0359	Protactinium	Pa
92		238.0289	Uranium	U
93	*	237	Neptunium	Np
94	*	244	Plutonium	Pu
95	*	243	Americium	Am
96	*	247	Curium	Cm
97	*	247	Berkelium	Bk
98	*	251	Californium	Cf
99	*	252	Einsteinium	Es
100	*	257	Fermium	Fm
101	*	258	Mendelevium	Md
102	*	259	Nobelium	No
103	*	262	Lawrencium	Lr